TRINITY

TRINITY

A GRAPHIC HISTORY OF THE FIRST ATOMIC BOMB

JONATHAN FETTER-VORM

A Novel Graphic from Hill and Wang
A division of Farrar, Straus and Giroux
New York

Hill and Wang
A division of Farrar, Straus and Giroux
18 West 18th Street, New York 10011

Printed in the United States of America
Published in 2012 by Hill and Wang
First paperback edition, 2013

The Library of Congress has cataloged the hardcover edition as follows:
Fetter-Vorm, Jonathan, 1983–
Trinity: a graphic history of the first atomic bomb / Jonathan Fetter-Vorm. — 1st ed.
p. cm.
ISBN 978-0-8090-9468-4 (cloth : alk. paper)
1. Atomic bomb—United States—History—Comic Books, strips, etc. 2. Manhattan Project (U.S.)—Comic books, strips, etc. I. Title.

QC773.3.U5 F47 2012
623.4′5119—dc23

2011036622

Paperback ISBN: 978-0-8090-9355-7

www.fsgbooks.com
www.twitter.com/fsgbooks • www.facebook.com/fsgbooks

19 20

When the stars threw down their spears,
And watered heaven with their tears,
Did he smile his work to see?
Did he who made the Lamb make thee?

—William Blake

It worked.

—J. Robert Oppenheimer

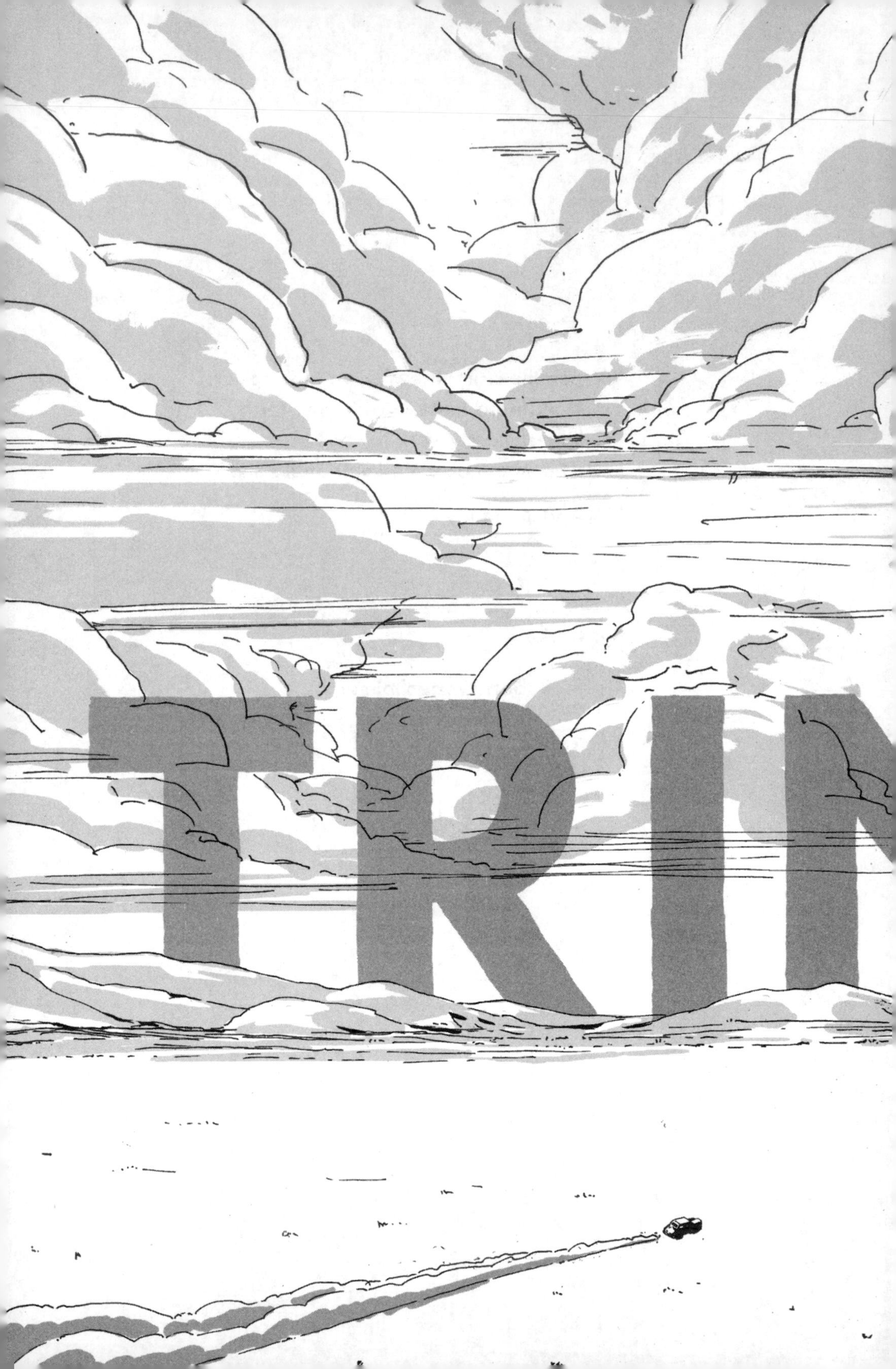

TRIN

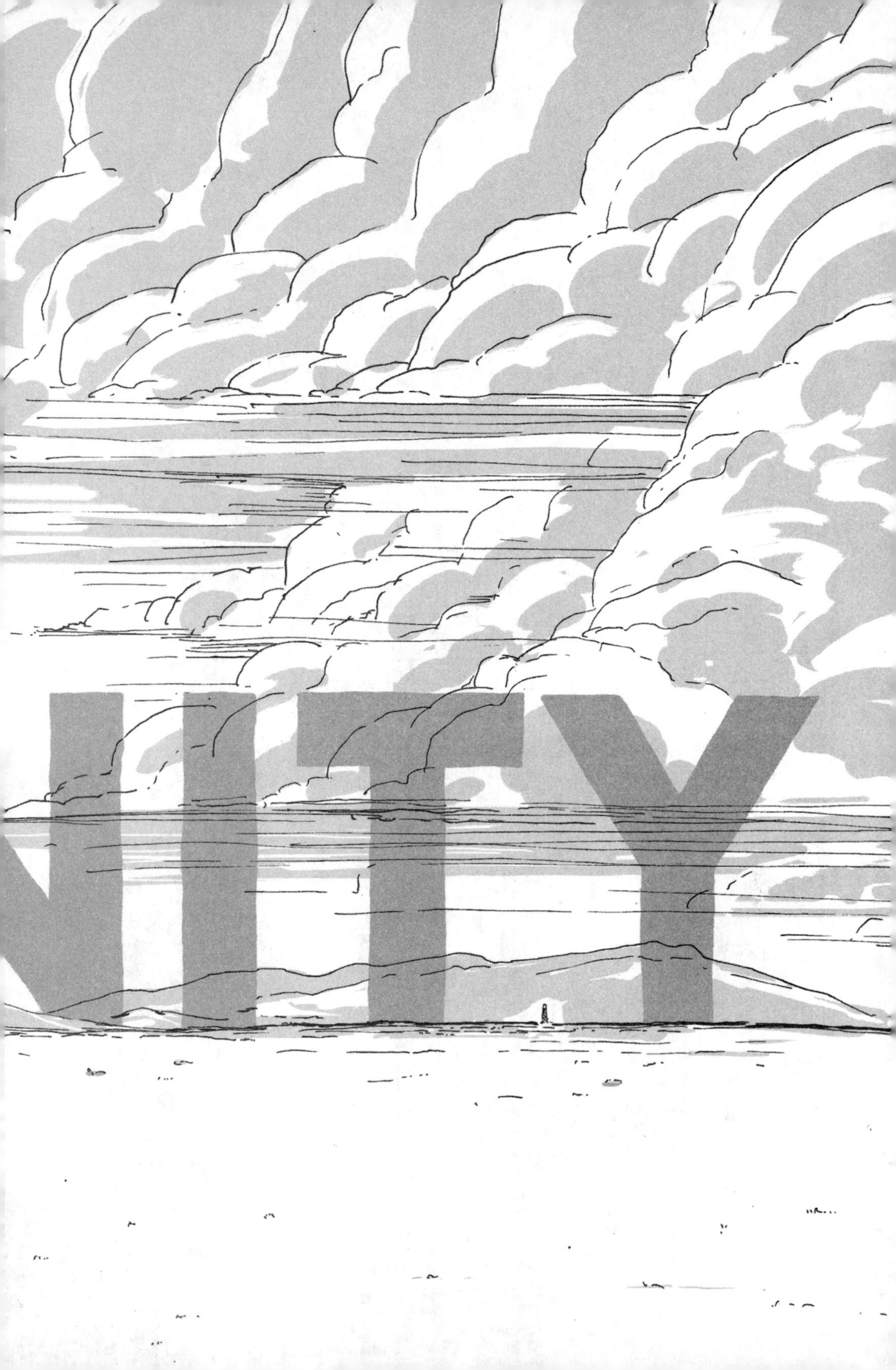
ITY

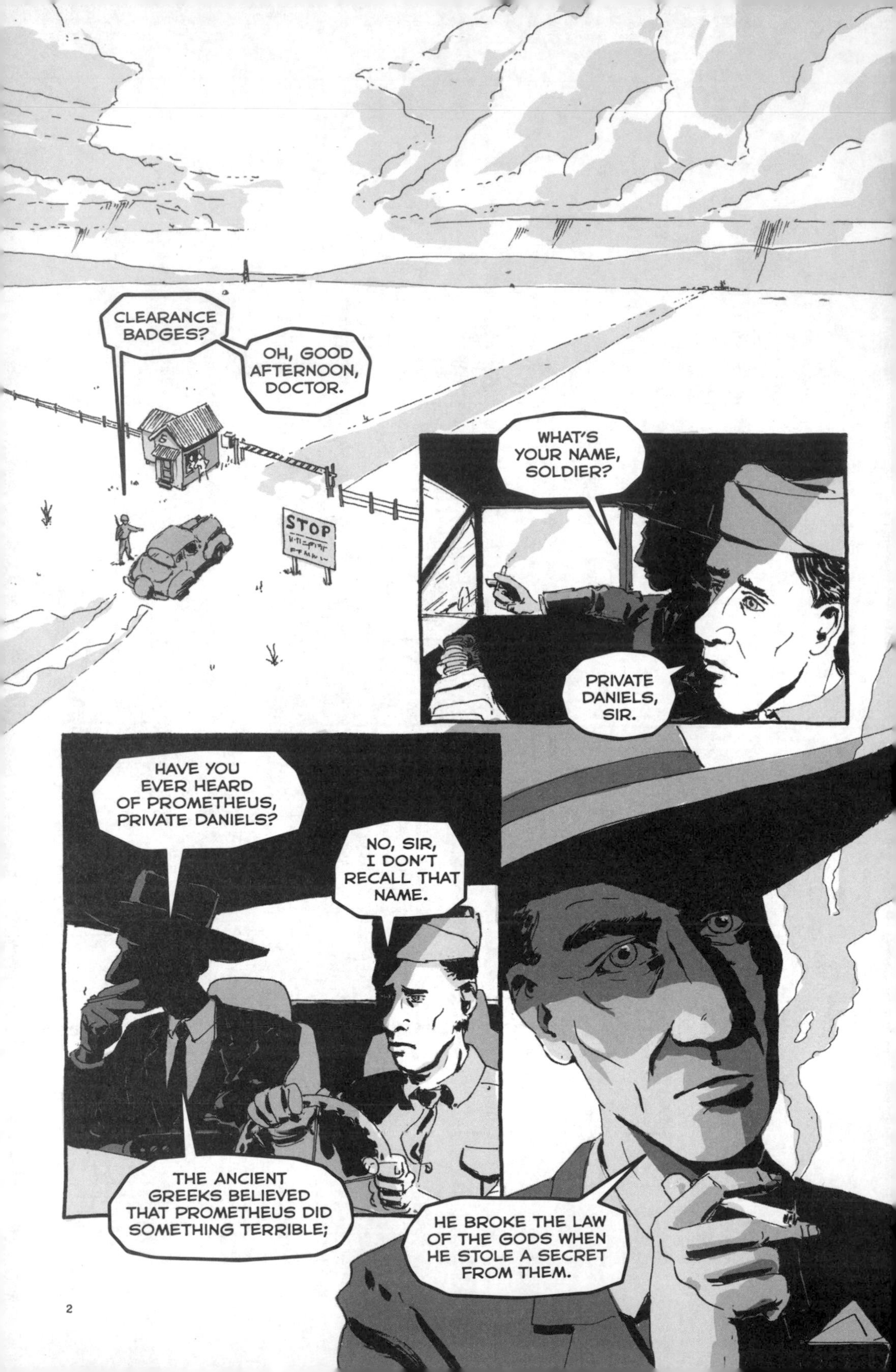
CLEARANCE BADGES?
OH, GOOD AFTERNOON, DOCTOR.
STOP
WHAT'S YOUR NAME, SOLDIER?
PRIVATE DANIELS, SIR.
HAVE YOU EVER HEARD OF PROMETHEUS, PRIVATE DANIELS?
NO, SIR, I DON'T RECALL THAT NAME.
THE ANCIENT GREEKS BELIEVED THAT PROMETHEUS DID SOMETHING TERRIBLE;
HE BROKE THE LAW OF THE GODS WHEN HE STOLE A SECRET FROM THEM.

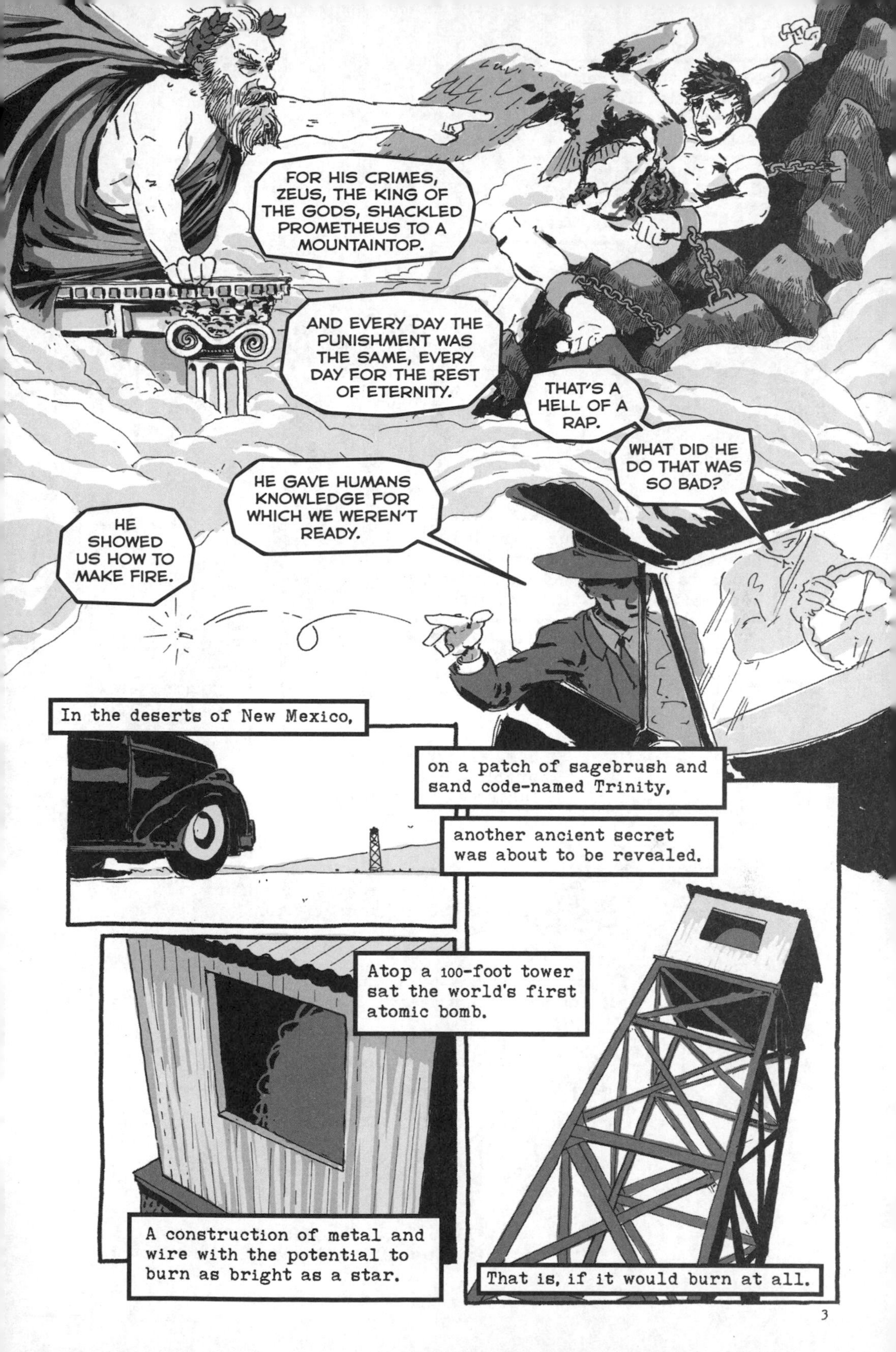

FOR HIS CRIMES, ZEUS, THE KING OF THE GODS, SHACKLED PROMETHEUS TO A MOUNTAINTOP.
AND EVERY DAY THE PUNISHMENT WAS THE SAME, EVERY DAY FOR THE REST OF ETERNITY.
THAT'S A HELL OF A RAP.
WHAT DID HE DO THAT WAS SO BAD?
HE GAVE HUMANS KNOWLEDGE FOR WHICH WE WEREN'T READY.
HE SHOWED US HOW TO MAKE FIRE.
In the deserts of New Mexico,
on a patch of sagebrush and sand code-named Trinity,
another ancient secret was about to be revealed.
Atop a 100-foot tower sat the world's first atomic bomb.
A construction of metal and wire with the potential to burn as bright as a star.
That is, if it would burn at all.

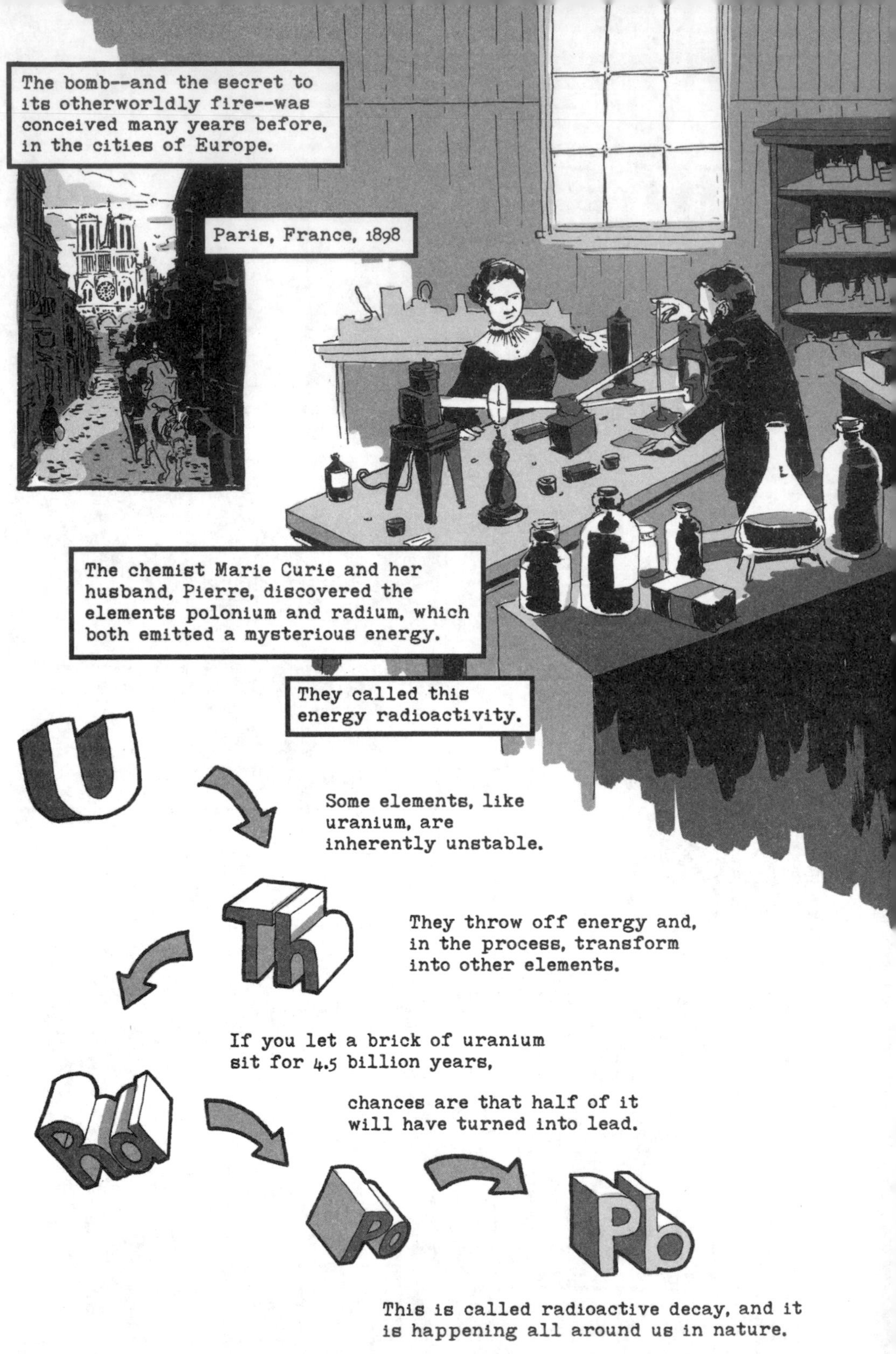
The bomb--and the secret to its otherworldly fire--was conceived many years before, in the cities of Europe.
Paris, France, 1898
The chemist Marie Curie and her husband, Pierre, discovered the elements polonium and radium, which both emitted a mysterious energy.
They called this energy radioactivity.
U
Some elements, like uranium, are inherently unstable.
Th
They throw off energy and, in the process, transform into other elements.
If you let a brick of uranium sit for 4.5 billion years,
Ra
chances are that half of it will have turned into lead.
Po
Pb
This is called radioactive decay, and it is happening all around us in nature.

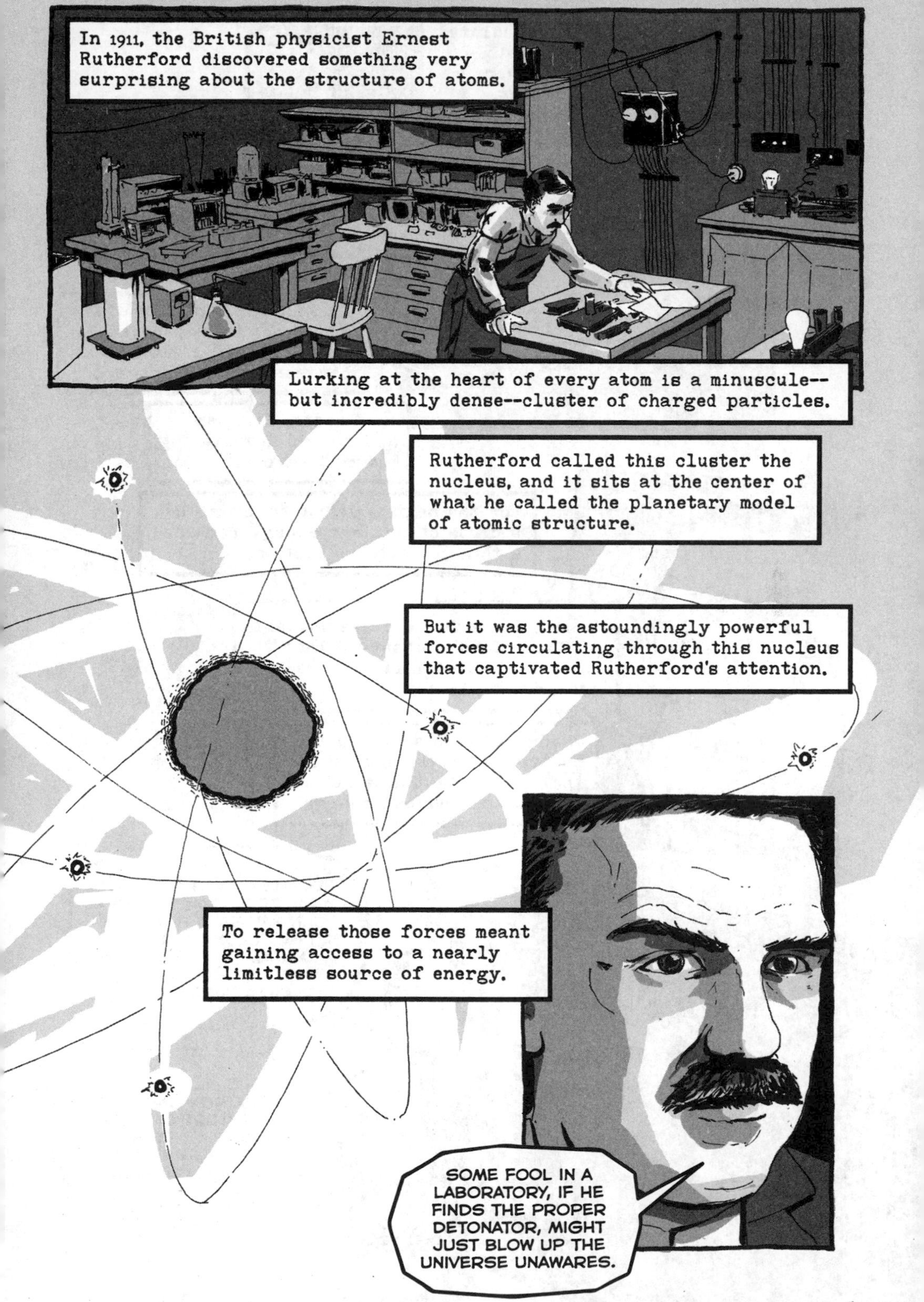

In 1911, the British physicist Ernest Rutherford discovered something very surprising about the structure of atoms.
Lurking at the heart of every atom is a minuscule--but incredibly dense--cluster of charged particles.
Rutherford called this cluster the nucleus, and it sits at the center of what he called the planetary model of atomic structure.
But it was the astoundingly powerful forces circulating through this nucleus that captivated Rutherford's attention.
To release those forces meant gaining access to a nearly limitless source of energy.
SOME FOOL IN A LABORATORY, IF HE FINDS THE PROPER DETONATOR, MIGHT JUST BLOW UP THE UNIVERSE UNAWARES.

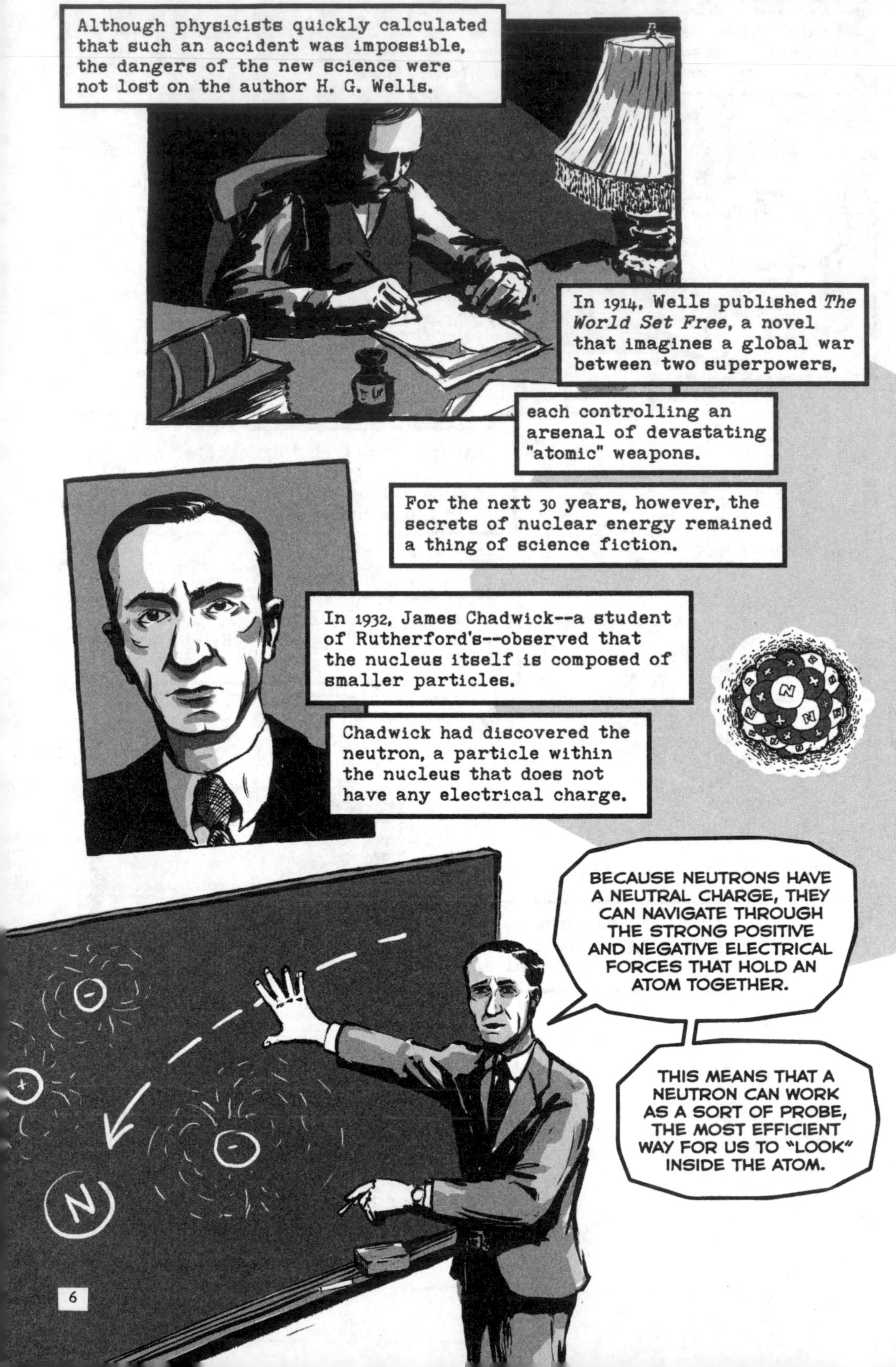
Although physicists quickly calculated that such an accident was impossible, the dangers of the new science were not lost on the author H. G. Wells.
In 1914, Wells published *The World Set Free*, a novel that imagines a global war between two superpowers,
each controlling an arsenal of devastating "atomic" weapons.
For the next 30 years, however, the secrets of nuclear energy remained a thing of science fiction.
In 1932, James Chadwick--a student of Rutherford's--observed that the nucleus itself is composed of smaller particles.
Chadwick had discovered the neutron, a particle within the nucleus that does not have any electrical charge.
BECAUSE NEUTRONS HAVE A NEUTRAL CHARGE, THEY CAN NAVIGATE THROUGH THE STRONG POSITIVE AND NEGATIVE ELECTRICAL FORCES THAT HOLD AN ATOM TOGETHER.
THIS MEANS THAT A NEUTRON CAN WORK AS A SORT OF PROBE, THE MOST EFFICIENT WAY FOR US TO "LOOK" INSIDE THE ATOM.

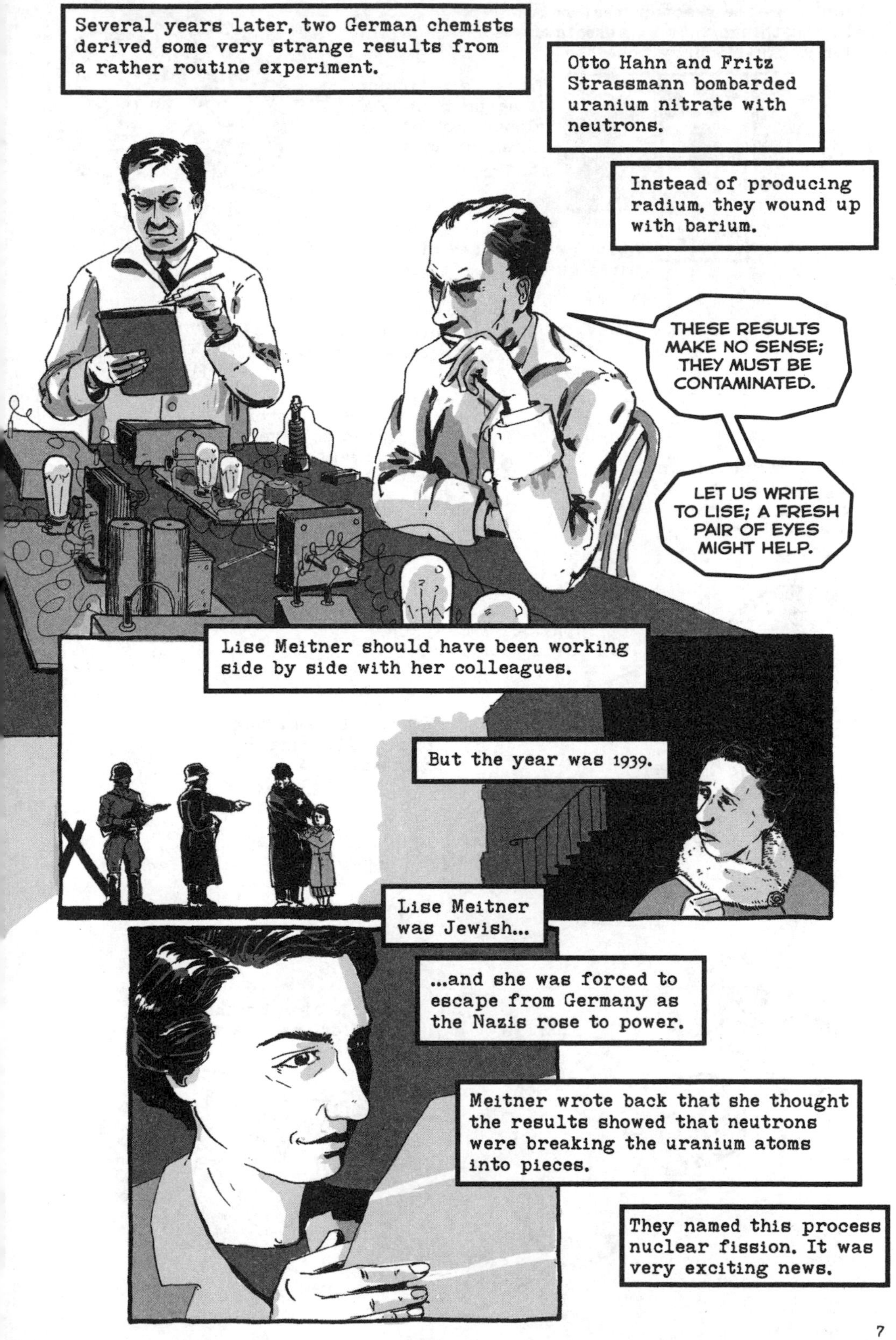
Several years later, two German chemists derived some very strange results from a rather routine experiment.
Otto Hahn and Fritz Strassmann bombarded uranium nitrate with neutrons.
Instead of producing radium, they wound up with barium.
THESE RESULTS MAKE NO SENSE; THEY MUST BE CONTAMINATED.
LET US WRITE TO LISE; A FRESH PAIR OF EYES MIGHT HELP.
Lise Meitner should have been working side by side with her colleagues.
But the year was 1939.
Lise Meitner was Jewish...
...and she was forced to escape from Germany as the Nazis rose to power.
Meitner wrote back that she thought the results showed that neutrons were breaking the uranium atoms into pieces.
They named this process nuclear fission. It was very exciting news.

Fission is the reaction that makes all the things that we associate with nuclear energy possible.

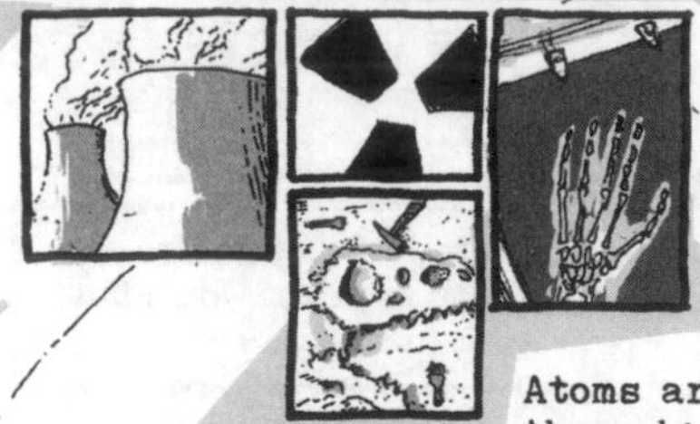

To explain fission, it helps to take a closer look at the inside of an atom.

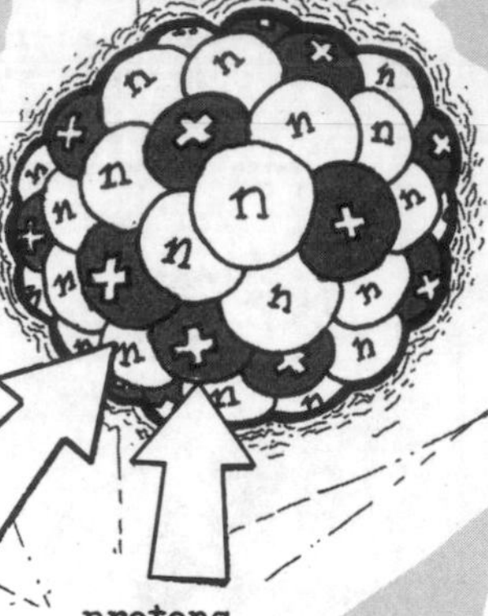

Atoms are composed of three kinds of particles:

neutrons (which, as Chadwick discovered, carry a neutral charge),

protons (which are positively charged),

and electrons (which have a negative charge).

These particles bind together thanks to the interplay of very powerful opposing forces.

The electrostatic force pushes particles away from each other.

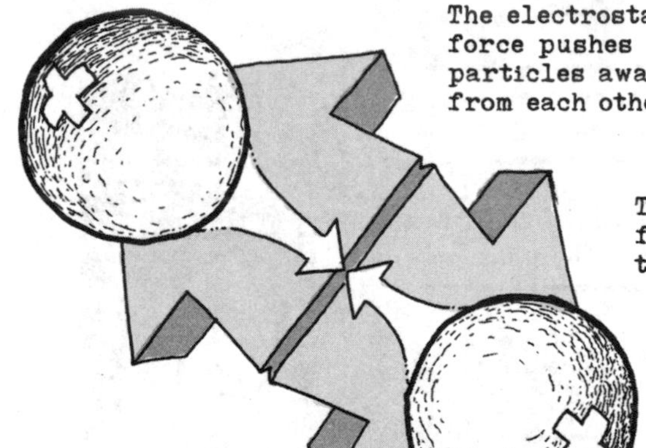

The so-called strong force pulls particles together.

When these powerful forces fall out of equilibrium,

the atom breaks, or fissions, apart.

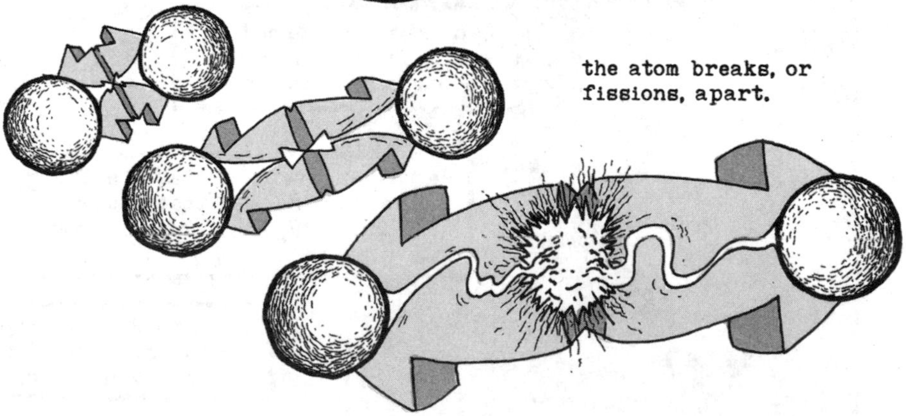

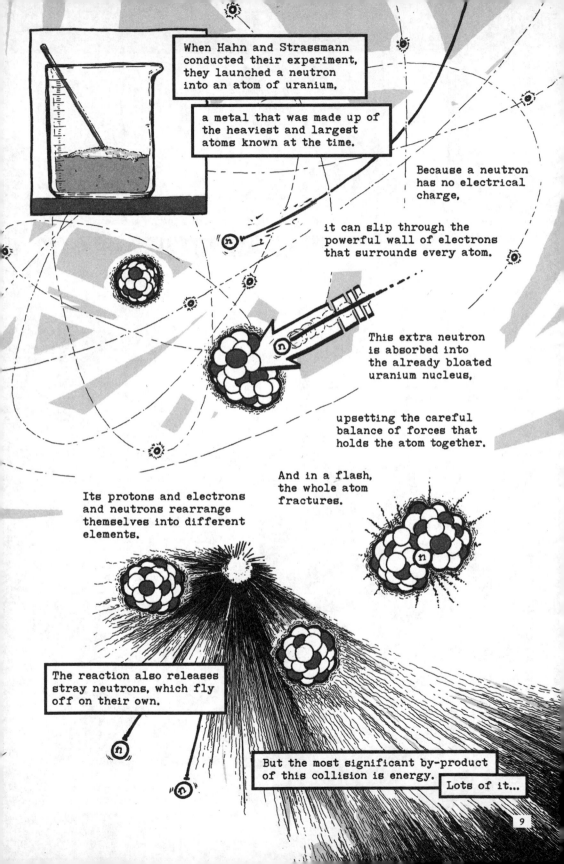

When Hahn and Strassmann conducted their experiment, they launched a neutron into an atom of uranium,

a metal that was made up of the heaviest and largest atoms known at the time.

Because a neutron has no electrical charge,

it can slip through the powerful wall of electrons that surrounds every atom.

This extra neutron is absorbed into the already bloated uranium nucleus,

upsetting the careful balance of forces that holds the atom together.

And in a flash, the whole atom fractures.

Its protons and electrons and neutrons rearrange themselves into different elements.

The reaction also releases stray neutrons, which fly off on their own.

But the most significant by-product of this collision is energy.

Lots of it...

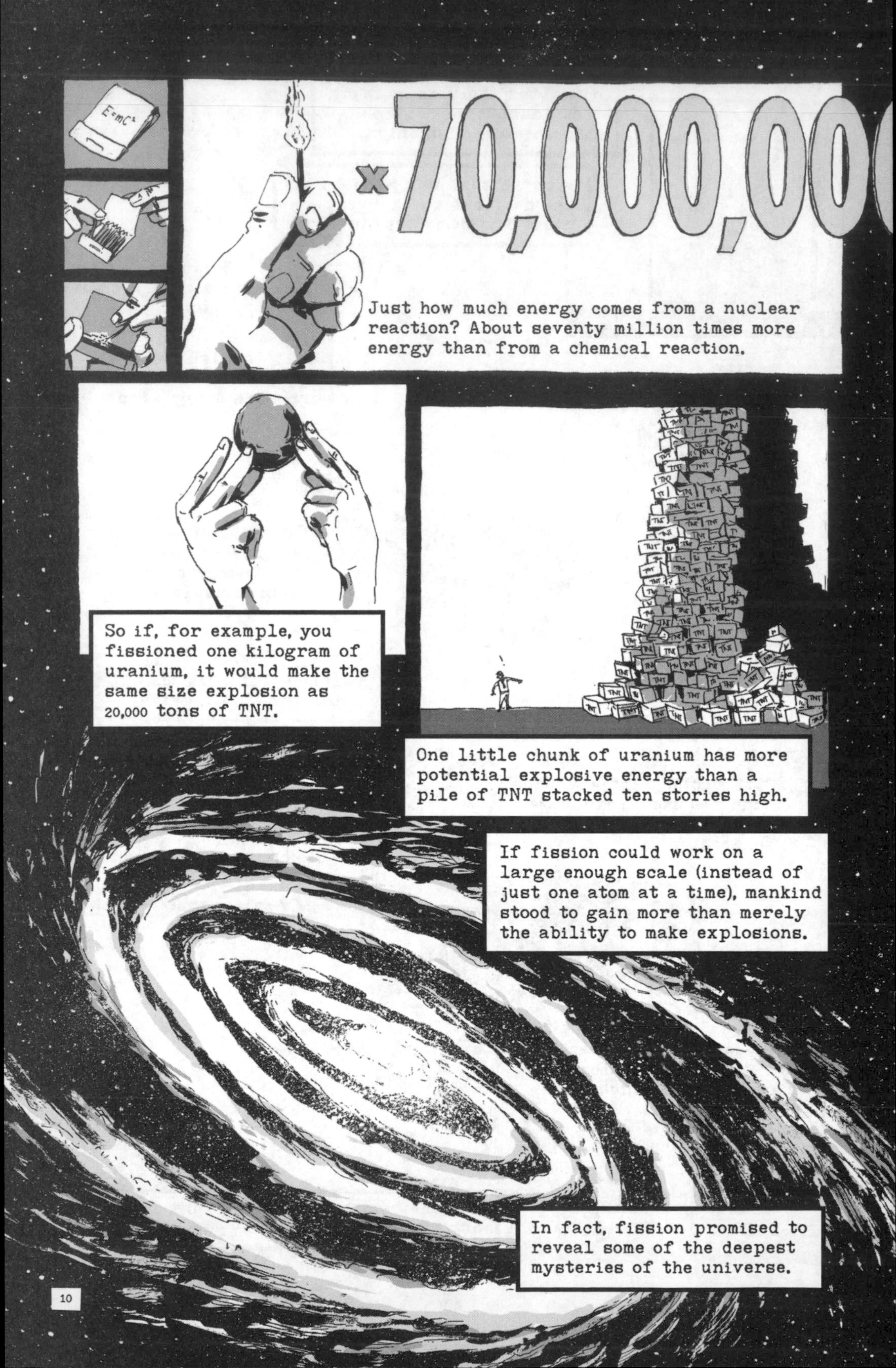
E=mC²
x 70,000,00
Just how much energy comes from a nuclear reaction? About seventy million times more energy than from a chemical reaction.
So if, for example, you fissioned one kilogram of uranium, it would make the same size explosion as 20,000 tons of TNT.
TNT
One little chunk of uranium has more potential explosive energy than a pile of TNT stacked ten stories high.
If fission could work on a large enough scale (instead of just one atom at a time), mankind stood to gain more than merely the ability to make explosions.
In fact, fission promised to reveal some of the deepest mysteries of the universe.

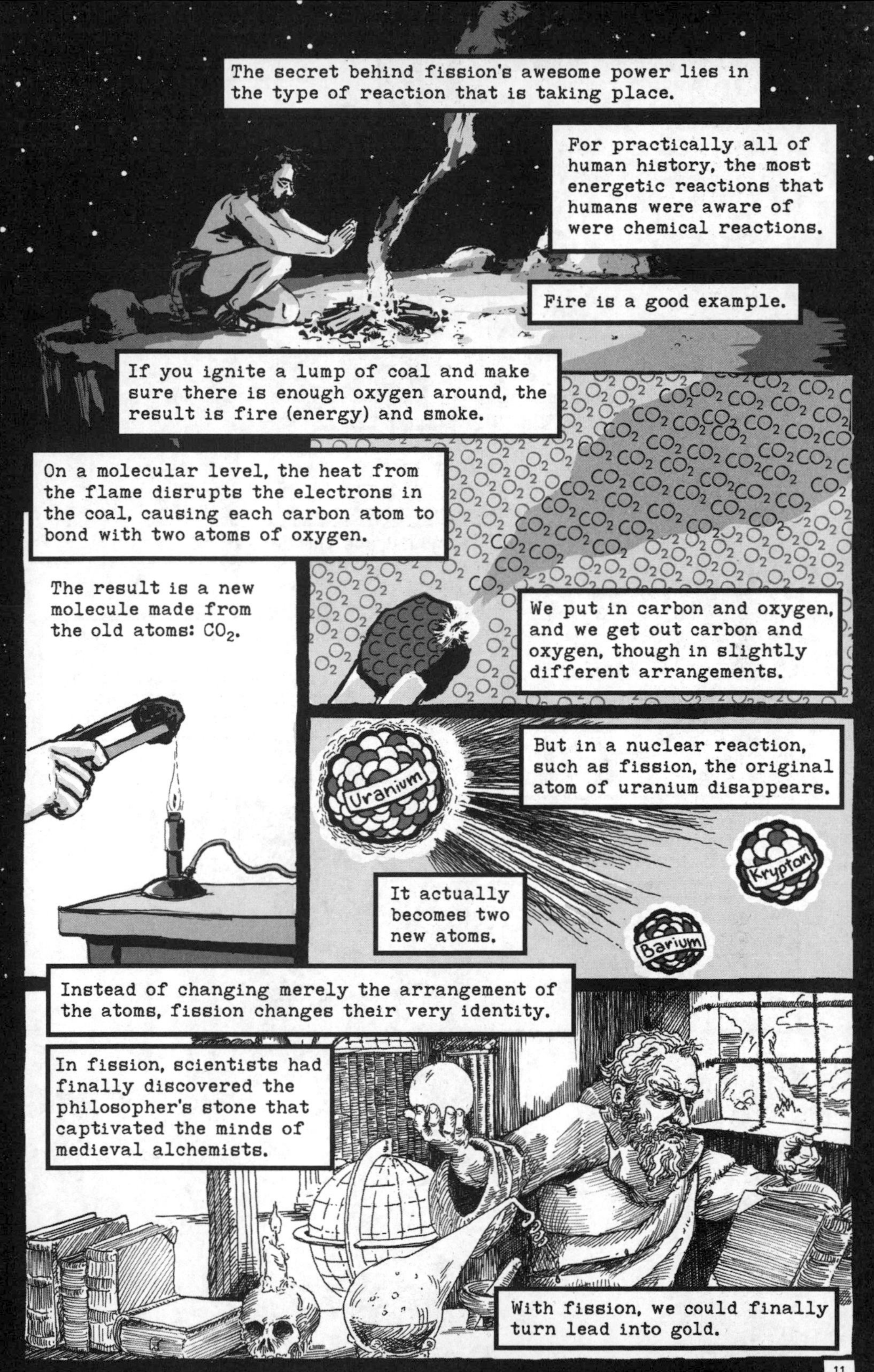
The secret behind fission's awesome power lies in the type of reaction that is taking place.
For practically all of human history, the most energetic reactions that humans were aware of were chemical reactions.
Fire is a good example.
If you ignite a lump of coal and make sure there is enough oxygen around, the result is fire (energy) and smoke.
On a molecular level, the heat from the flame disrupts the electrons in the coal, causing each carbon atom to bond with two atoms of oxygen.
The result is a new molecule made from the old atoms: CO_2.
We put in carbon and oxygen, and we get out carbon and oxygen, though in slightly different arrangements.
But in a nuclear reaction, such as fission, the original atom of uranium disappears.
Uranium
It actually becomes two new atoms.
Krypton
Barium
Instead of changing merely the arrangement of the atoms, fission changes their very identity.
In fission, scientists had finally discovered the philosopher's stone that captivated the minds of medieval alchemists.
With fission, we could finally turn lead into gold.

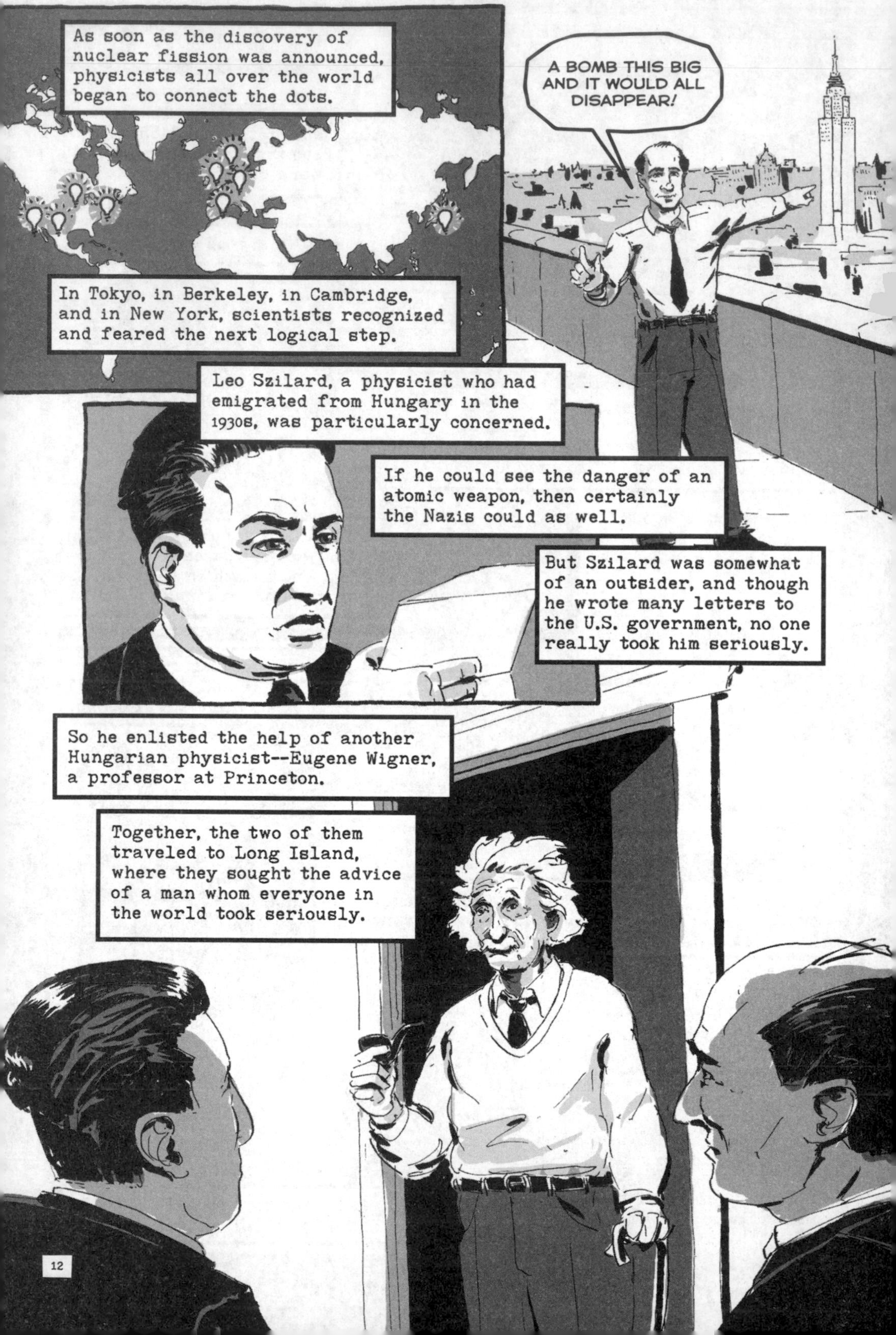
As soon as the discovery of nuclear fission was announced, physicists all over the world began to connect the dots.
A BOMB THIS BIG AND IT WOULD ALL DISAPPEAR!
In Tokyo, in Berkeley, in Cambridge, and in New York, scientists recognized and feared the next logical step.
Leo Szilard, a physicist who had emigrated from Hungary in the 1930s, was particularly concerned.
If he could see the danger of an atomic weapon, then certainly the Nazis could as well.
But Szilard was somewhat of an outsider, and though he wrote many letters to the U.S. government, no one really took him seriously.
So he enlisted the help of another Hungarian physicist--Eugene Wigner, a professor at Princeton.
Together, the two of them traveled to Long Island, where they sought the advice of a man whom everyone in the world took seriously.

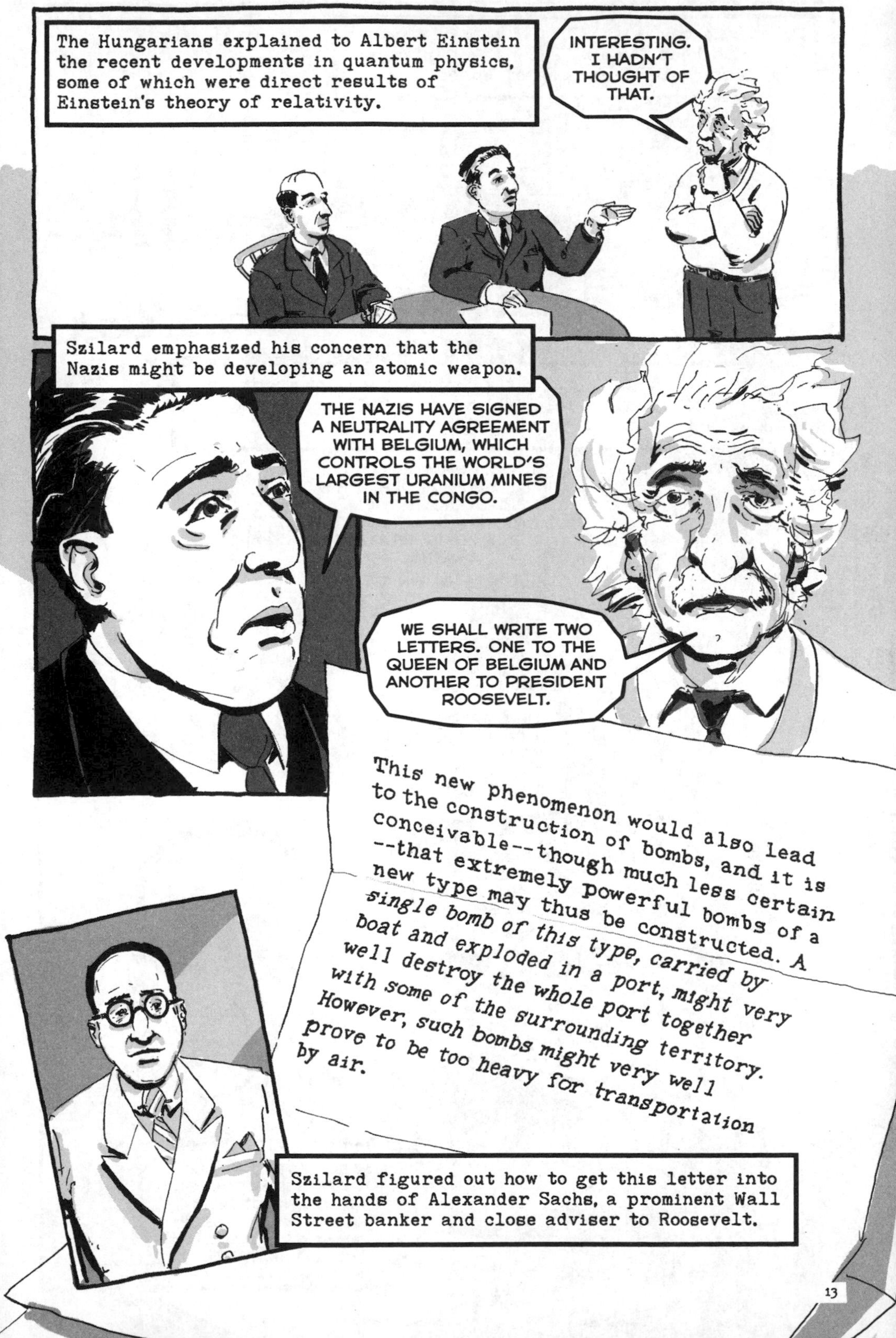
The Hungarians explained to Albert Einstein the recent developments in quantum physics, some of which were direct results of Einstein's theory of relativity.
INTERESTING. I HADN'T THOUGHT OF THAT.
Szilard emphasized his concern that the Nazis might be developing an atomic weapon.
THE NAZIS HAVE SIGNED A NEUTRALITY AGREEMENT WITH BELGIUM, WHICH CONTROLS THE WORLD'S LARGEST URANIUM MINES IN THE CONGO.
WE SHALL WRITE TWO LETTERS. ONE TO THE QUEEN OF BELGIUM AND ANOTHER TO PRESIDENT ROOSEVELT.
This new phenomenon would also lead to the construction of bombs, and it is conceivable--though much less certain --that extremely powerful bombs of a new type may thus be constructed. A single bomb of this type, carried by boat and exploded in a port, might very well destroy the whole port together with some of the surrounding territory. However, such bombs might very well prove to be too heavy for transportation by air.
Szilard figured out how to get this letter into the hands of Alexander Sachs, a prominent Wall Street banker and close adviser to Roosevelt.

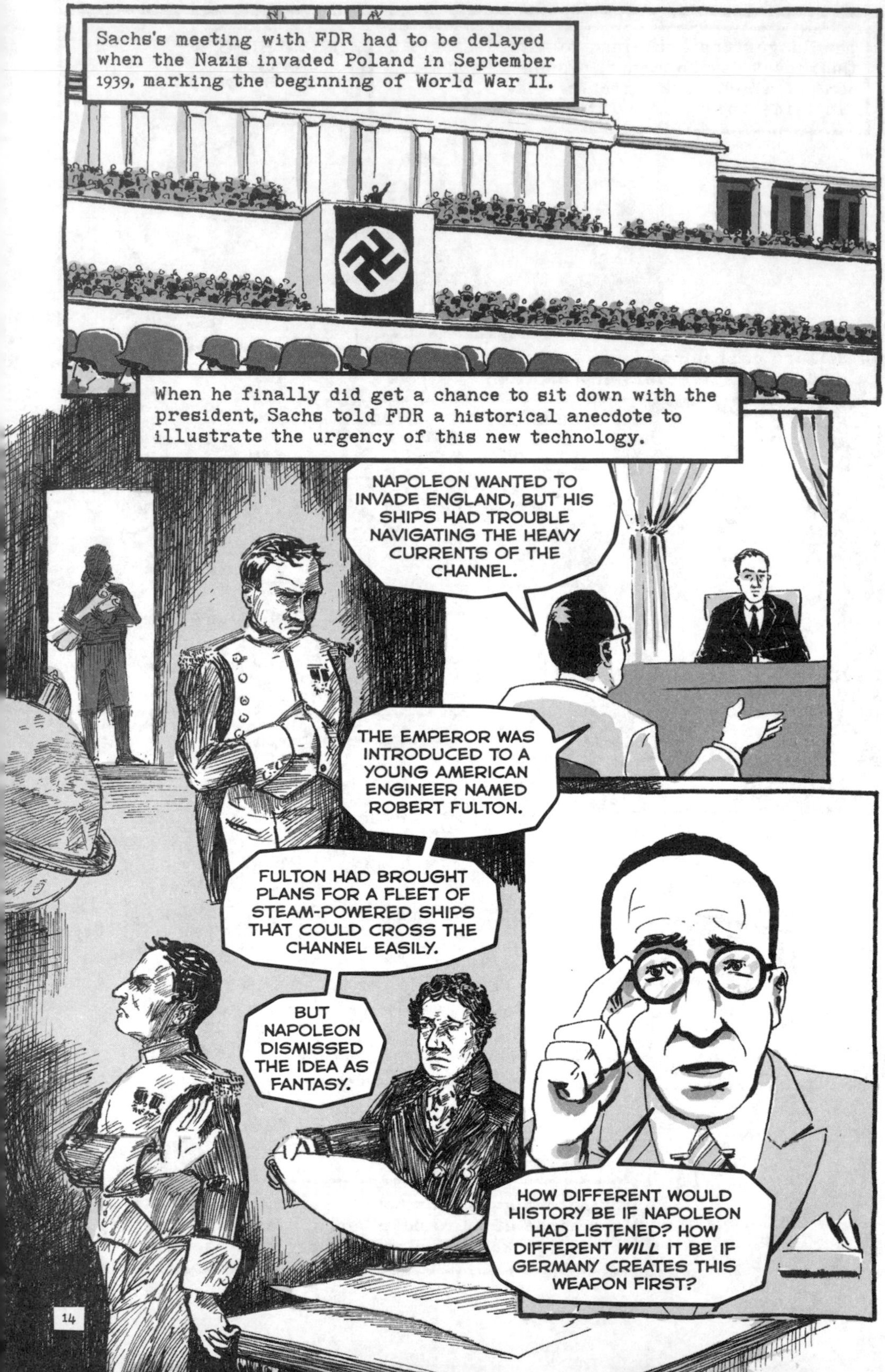
Sachs's meeting with FDR had to be delayed when the Nazis invaded Poland in September 1939, marking the beginning of World War II.
When he finally did get a chance to sit down with the president, Sachs told FDR a historical anecdote to illustrate the urgency of this new technology.
NAPOLEON WANTED TO INVADE ENGLAND, BUT HIS SHIPS HAD TROUBLE NAVIGATING THE HEAVY CURRENTS OF THE CHANNEL.
THE EMPEROR WAS INTRODUCED TO A YOUNG AMERICAN ENGINEER NAMED ROBERT FULTON.
FULTON HAD BROUGHT PLANS FOR A FLEET OF STEAM-POWERED SHIPS THAT COULD CROSS THE CHANNEL EASILY.
BUT NAPOLEON DISMISSED THE IDEA AS FANTASY.
HOW DIFFERENT WOULD HISTORY BE IF NAPOLEON HAD LISTENED? HOW DIFFERENT *WILL* IT BE IF GERMANY CREATES THIS WEAPON FIRST?

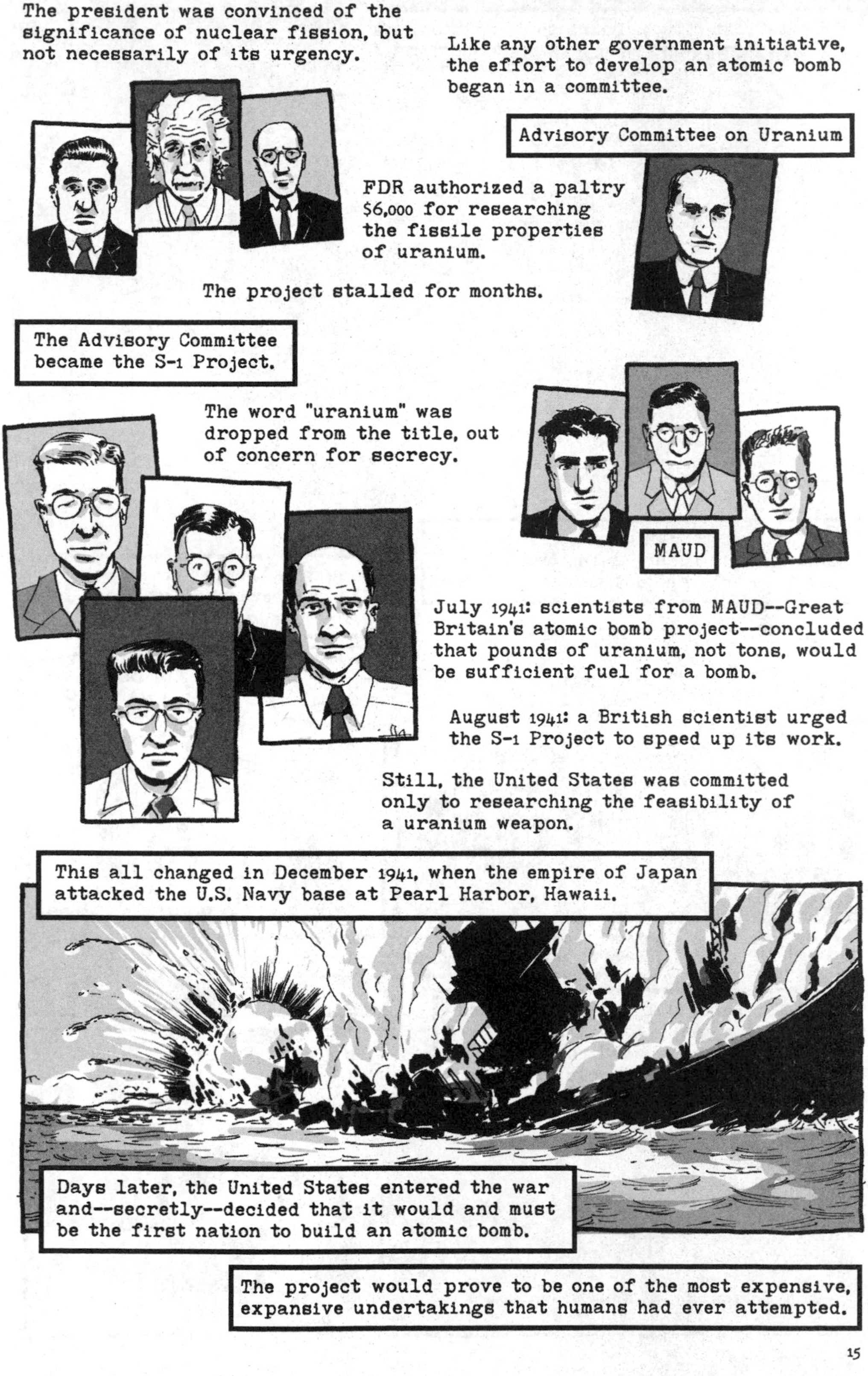
The president was convinced of the significance of nuclear fission, but not necessarily of its urgency.
Like any other government initiative, the effort to develop an atomic bomb began in a committee.
Advisory Committee on Uranium
FDR authorized a paltry $6,000 for researching the fissile properties of uranium.
The project stalled for months.
The Advisory Committee became the S-1 Project.
The word "uranium" was dropped from the title, out of concern for secrecy.
MAUD
July 1941: scientists from MAUD--Great Britain's atomic bomb project--concluded that pounds of uranium, not tons, would be sufficient fuel for a bomb.
August 1941: a British scientist urged the S-1 Project to speed up its work.
Still, the United States was committed only to researching the feasibility of a uranium weapon.
This all changed in December 1941, when the empire of Japan attacked the U.S. Navy base at Pearl Harbor, Hawaii.
Days later, the United States entered the war and--secretly--decided that it would and must be the first nation to build an atomic bomb.
The project would prove to be one of the most expensive, expansive undertakings that humans had ever attempted.

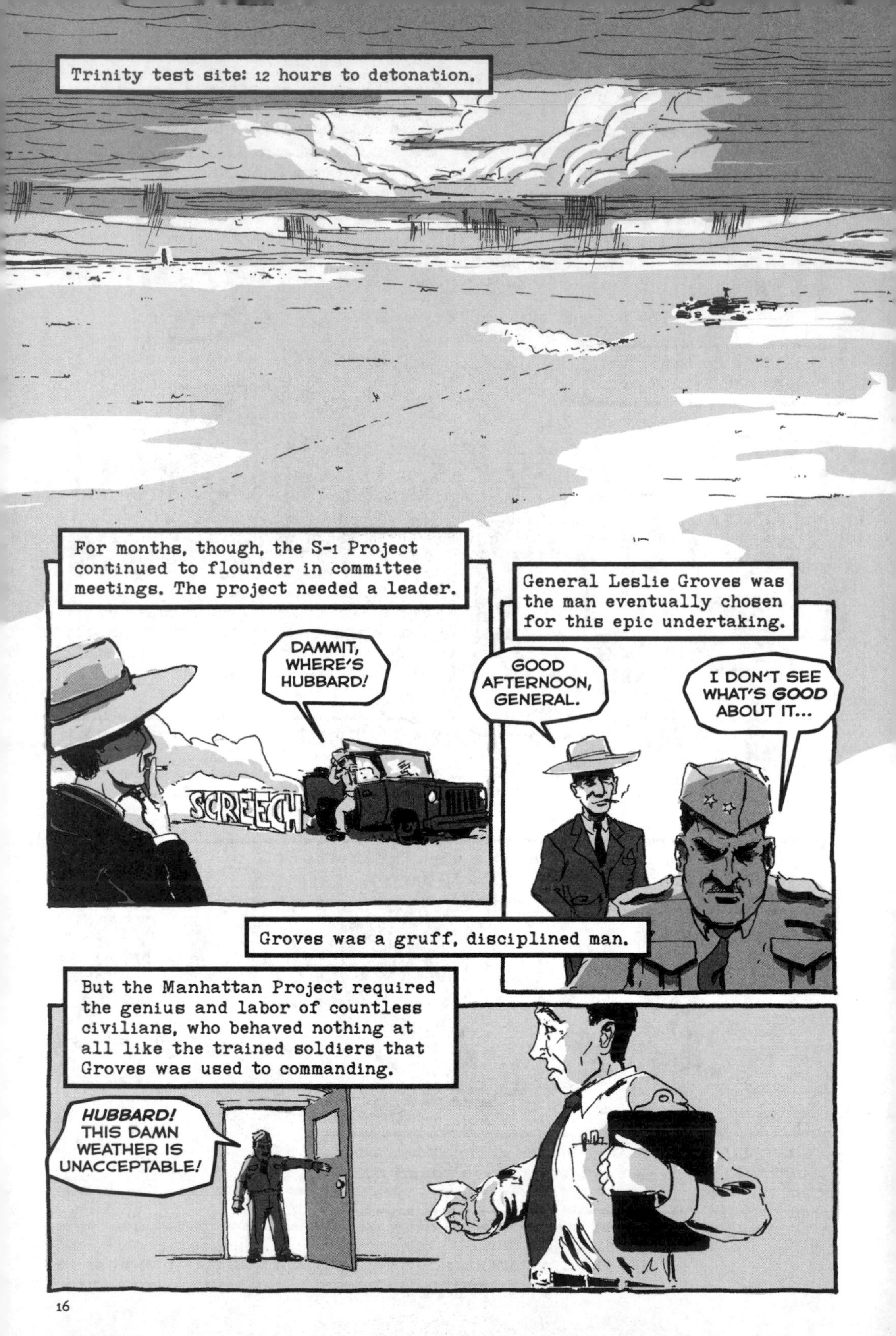
Trinity test site: 12 hours to detonation.
For months, though, the S-1 Project continued to flounder in committee meetings. The project needed a leader.
DAMMIT, WHERE'S HUBBARD!
SCREECH
General Leslie Groves was the man eventually chosen for this epic undertaking.
GOOD AFTERNOON, GENERAL.
I DON'T SEE WHAT'S *GOOD* ABOUT IT...
Groves was a gruff, disciplined man.
But the Manhattan Project required the genius and labor of countless civilians, who behaved nothing at all like the trained soldiers that Groves was used to commanding.
HUBBARD! THIS DAMN WEATHER IS UNACCEPTABLE!

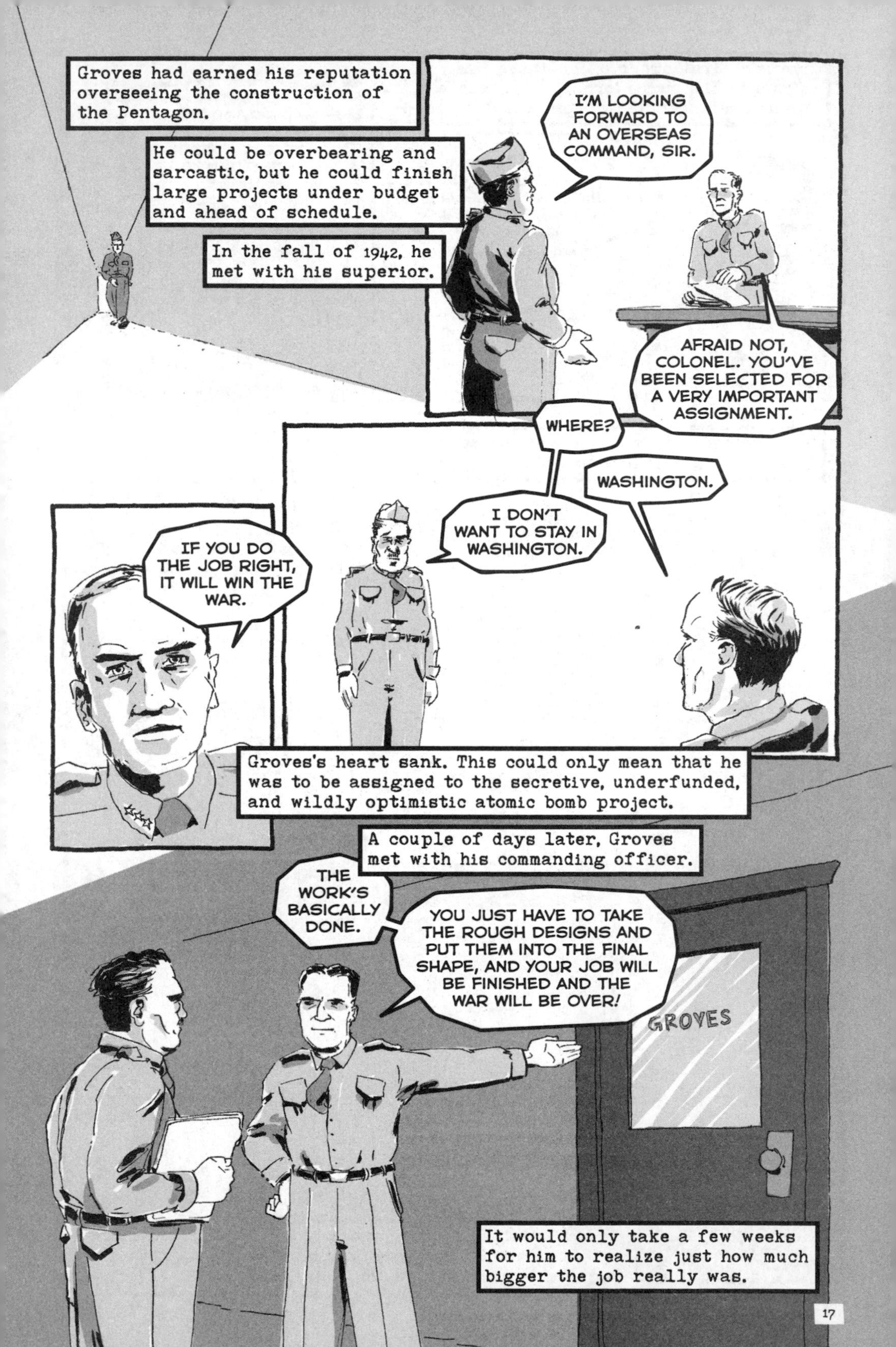

Groves had earned his reputation overseeing the construction of the Pentagon.
He could be overbearing and sarcastic, but he could finish large projects under budget and ahead of schedule.
In the fall of 1942, he met with his superior.
I'M LOOKING FORWARD TO AN OVERSEAS COMMAND, SIR.
AFRAID NOT, COLONEL. YOU'VE BEEN SELECTED FOR A VERY IMPORTANT ASSIGNMENT.
WHERE?
WASHINGTON.
I DON'T WANT TO STAY IN WASHINGTON.
IF YOU DO THE JOB RIGHT, IT WILL WIN THE WAR.
Groves's heart sank. This could only mean that he was to be assigned to the secretive, underfunded, and wildly optimistic atomic bomb project.
A couple of days later, Groves met with his commanding officer.
THE WORK'S BASICALLY DONE.
YOU JUST HAVE TO TAKE THE ROUGH DESIGNS AND PUT THEM INTO THE FINAL SHAPE, AND YOUR JOB WILL BE FINISHED AND THE WAR WILL BE OVER!
GROVES
It would only take a few weeks for him to realize just how much bigger the job really was.

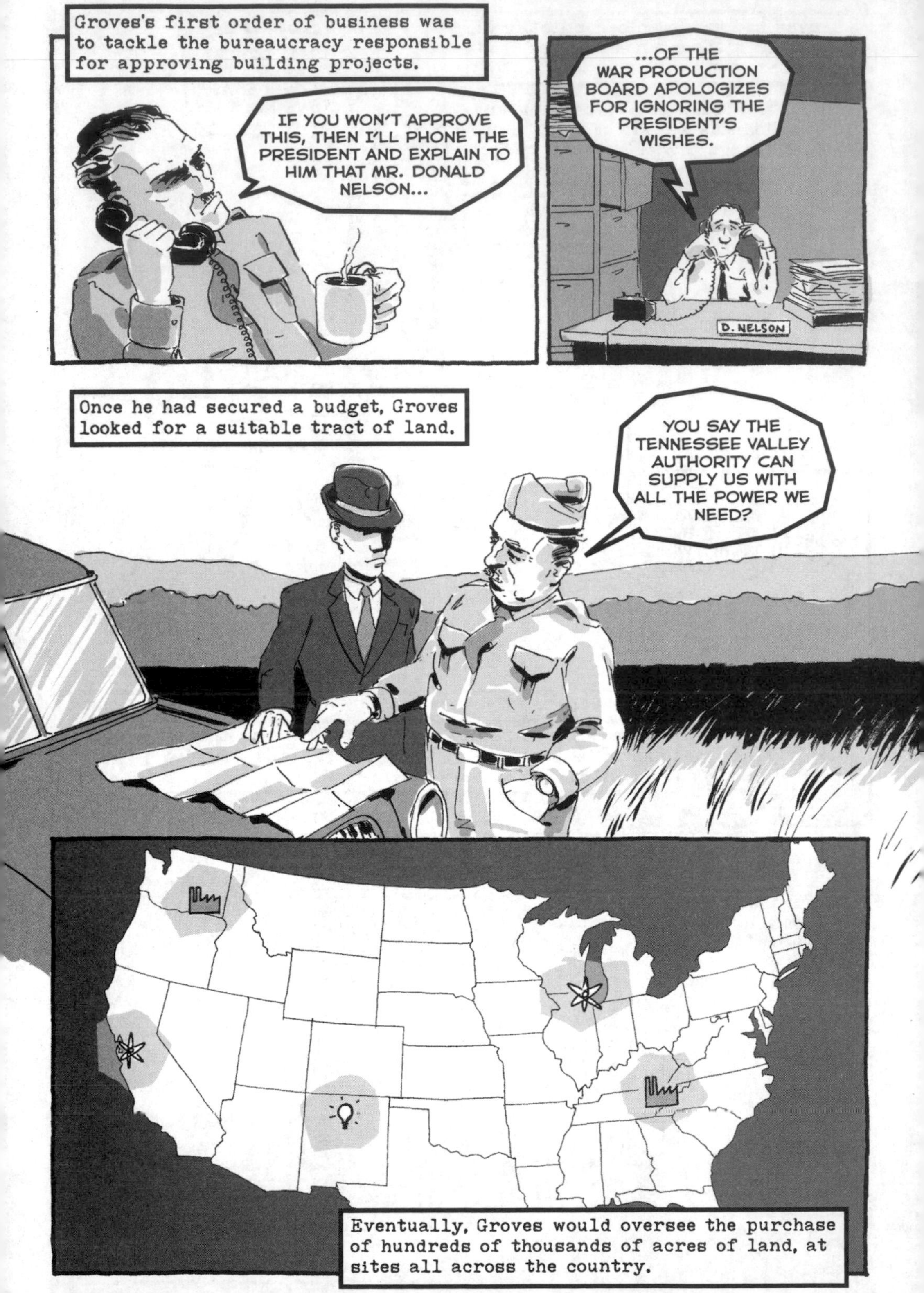
Groves's first order of business was to tackle the bureaucracy responsible for approving building projects.
IF YOU WON'T APPROVE THIS, THEN I'LL PHONE THE PRESIDENT AND EXPLAIN TO HIM THAT MR. DONALD NELSON...
...OF THE WAR PRODUCTION BOARD APOLOGIZES FOR IGNORING THE PRESIDENT'S WISHES.
D. NELSON
Once he had secured a budget, Groves looked for a suitable tract of land.
YOU SAY THE TENNESSEE VALLEY AUTHORITY CAN SUPPLY US WITH ALL THE POWER WE NEED?
Eventually, Groves would oversee the purchase of hundreds of thousands of acres of land, at sites all across the country.

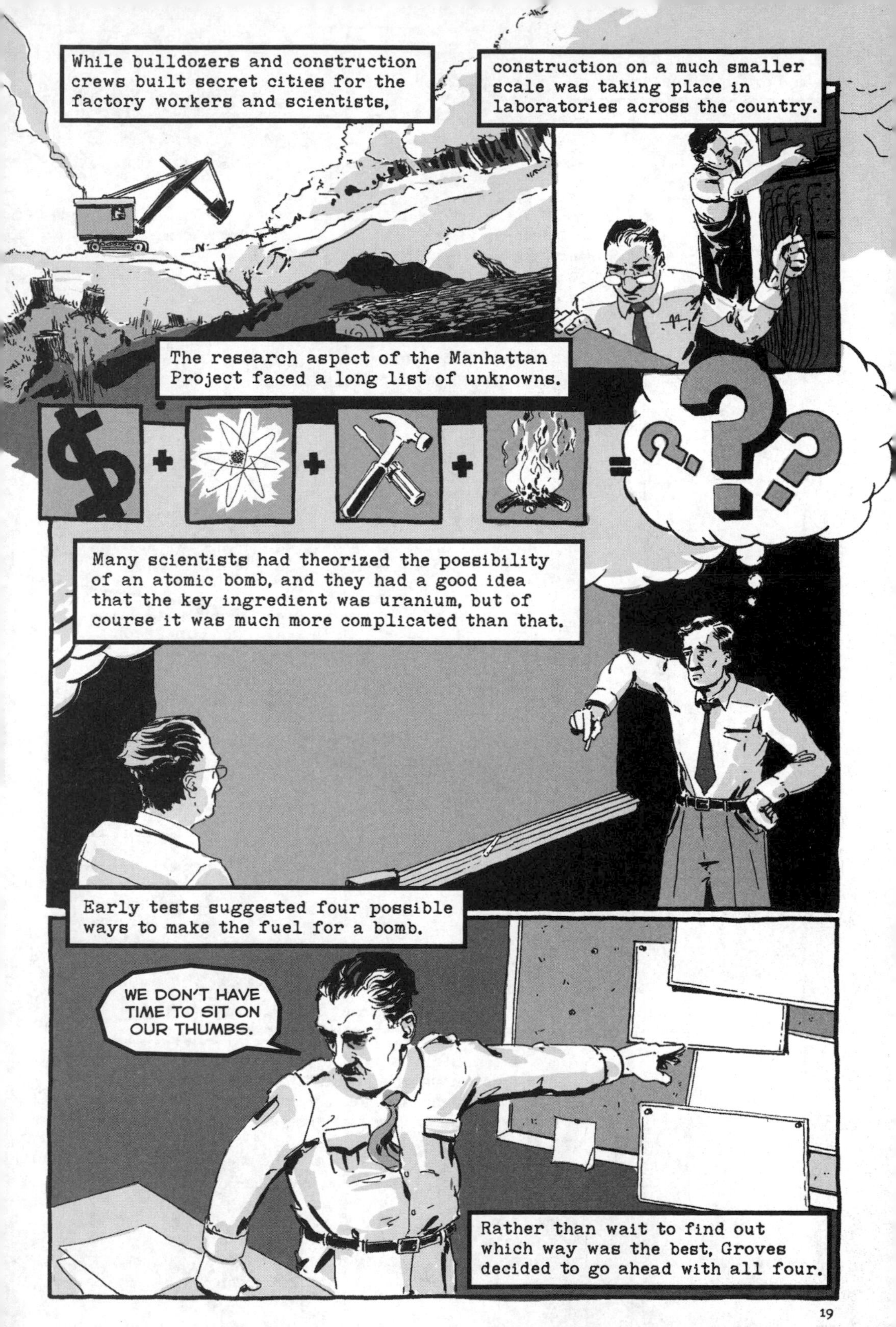
While bulldozers and construction crews built secret cities for the factory workers and scientists,
construction on a much smaller scale was taking place in laboratories across the country.
The research aspect of the Manhattan Project faced a long list of unknowns.
Many scientists had theorized the possibility of an atomic bomb, and they had a good idea that the key ingredient was uranium, but of course it was much more complicated than that.
Early tests suggested four possible ways to make the fuel for a bomb.
WE DON'T HAVE TIME TO SIT ON OUR THUMBS.
Rather than wait to find out which way was the best, Groves decided to go ahead with all four.

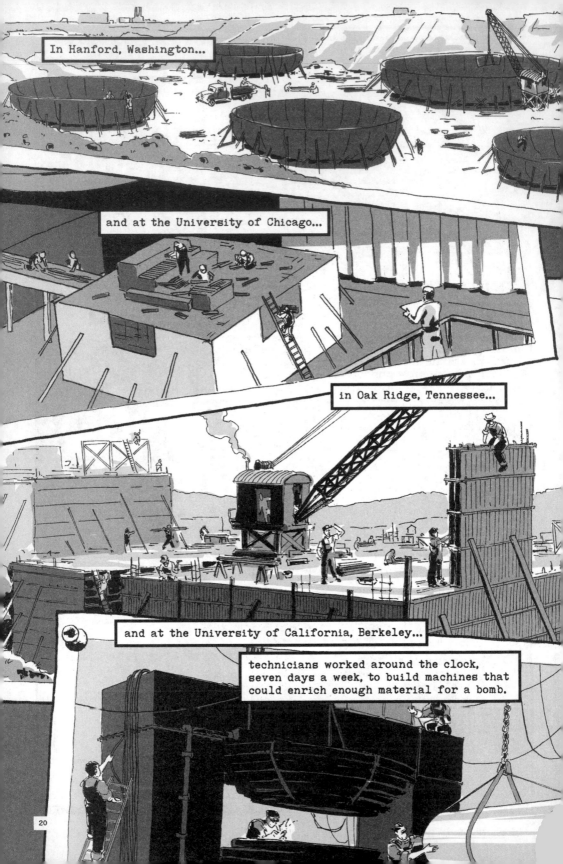

In Hanford, Washington...

and at the University of Chicago...

in Oak Ridge, Tennessee...

and at the University of California, Berkeley...

technicians worked around the clock, seven days a week, to build machines that could enrich enough material for a bomb.

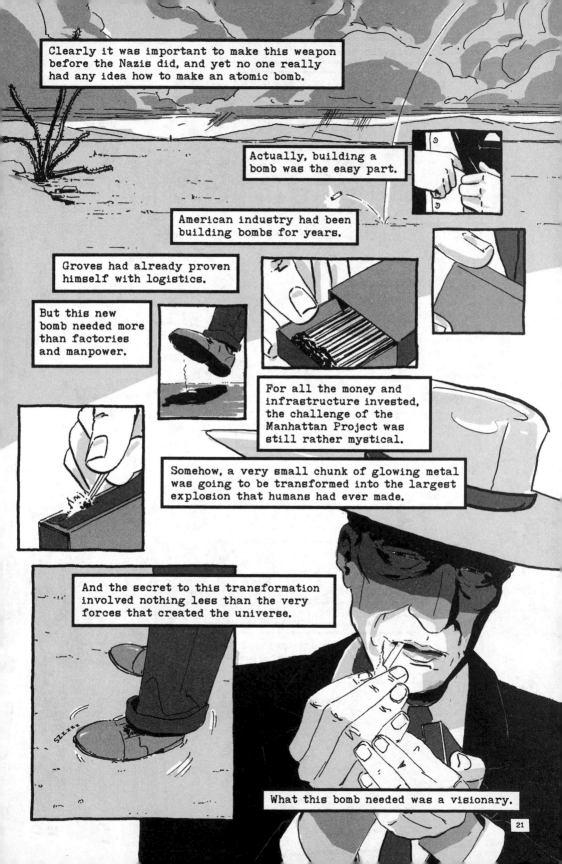

Clearly it was important to make this weapon before the Nazis did, and yet no one really had any idea how to make an atomic bomb.

Actually, building a bomb was the easy part.

American industry had been building bombs for years.

Groves had already proven himself with logistics.

But this new bomb needed more than factories and manpower.

For all the money and infrastructure invested, the challenge of the Manhattan Project was still rather mystical.

Somehow, a very small chunk of glowing metal was going to be transformed into the largest explosion that humans had ever made.

And the secret to this transformation involved nothing less than the very forces that created the universe.

What this bomb needed was a visionary.

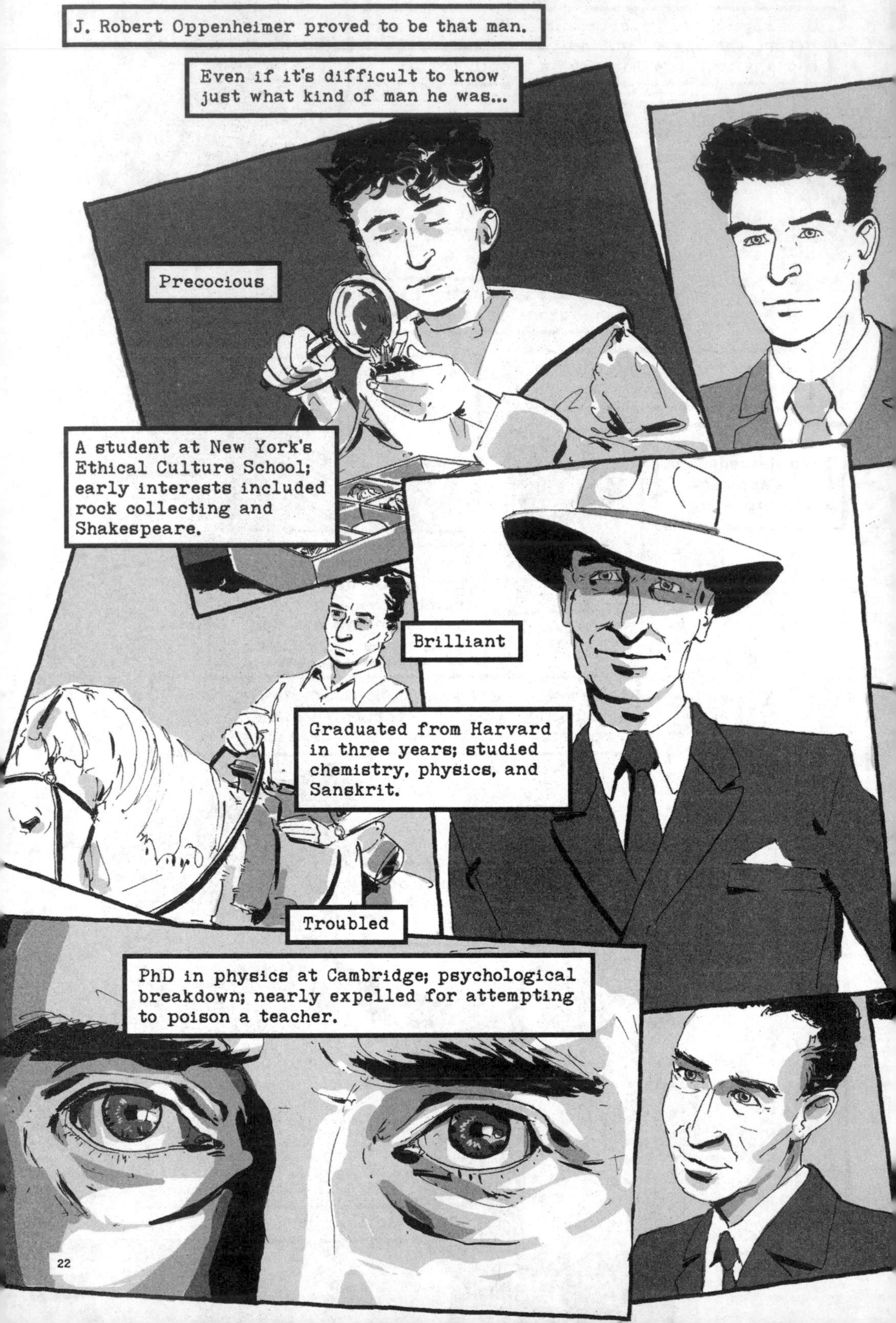
J. Robert Oppenheimer proved to be that man.
Even if it's difficult to know just what kind of man he was...
Precocious
A student at New York's Ethical Culture School; early interests included rock collecting and Shakespeare.
Brilliant
Graduated from Harvard in three years; studied chemistry, physics, and Sanskrit.
Troubled
PhD in physics at Cambridge; psychological breakdown; nearly expelled for attempting to poison a teacher.

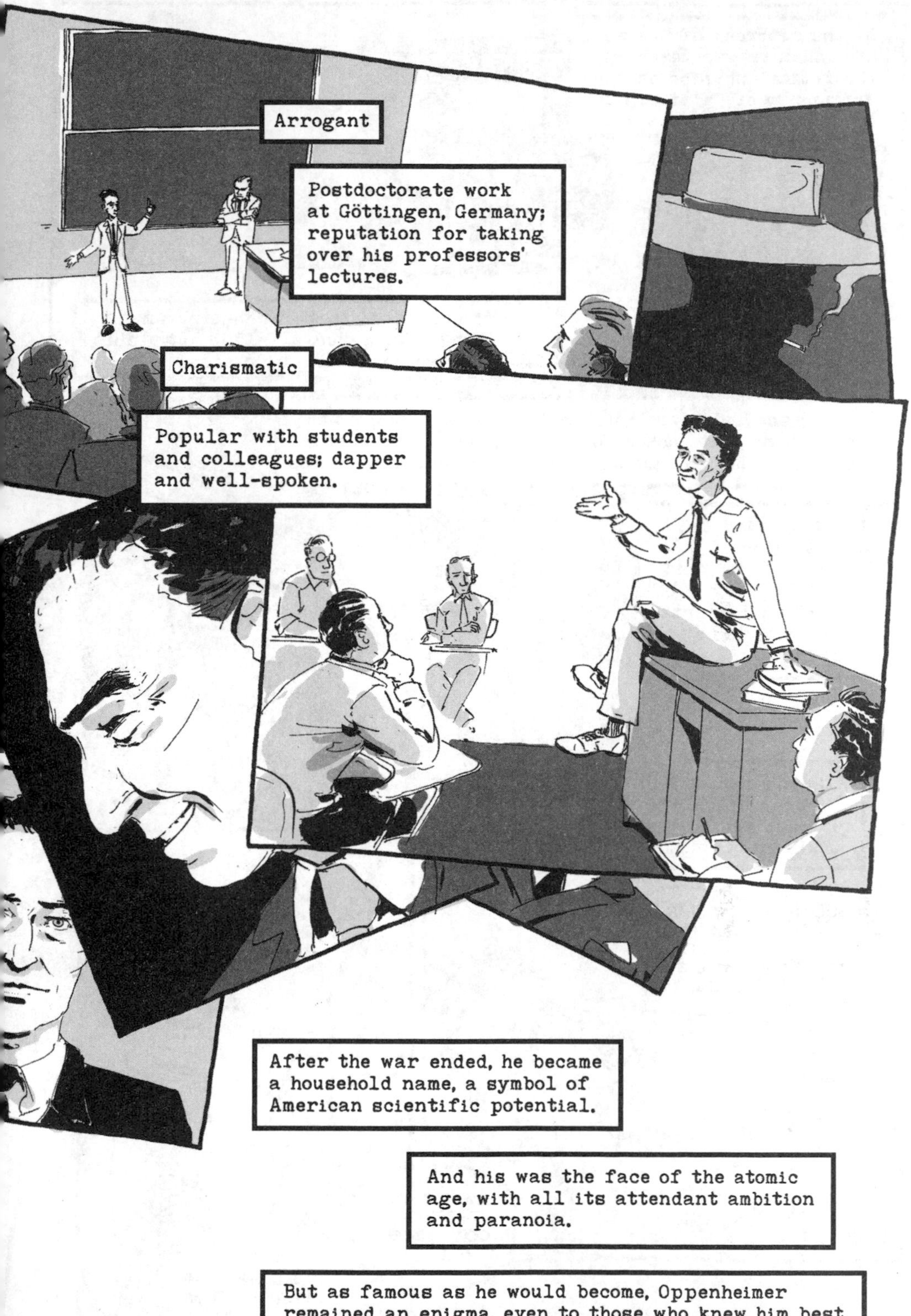
Arrogant
Postdoctorate work at Göttingen, Germany; reputation for taking over his professors' lectures.
Charismatic
Popular with students and colleagues; dapper and well-spoken.
After the war ended, he became a household name, a symbol of American scientific potential.
And his was the face of the atomic age, with all its attendant ambition and paranoia.
But as famous as he would become, Oppenheimer remained an enigma, even to those who knew him best.

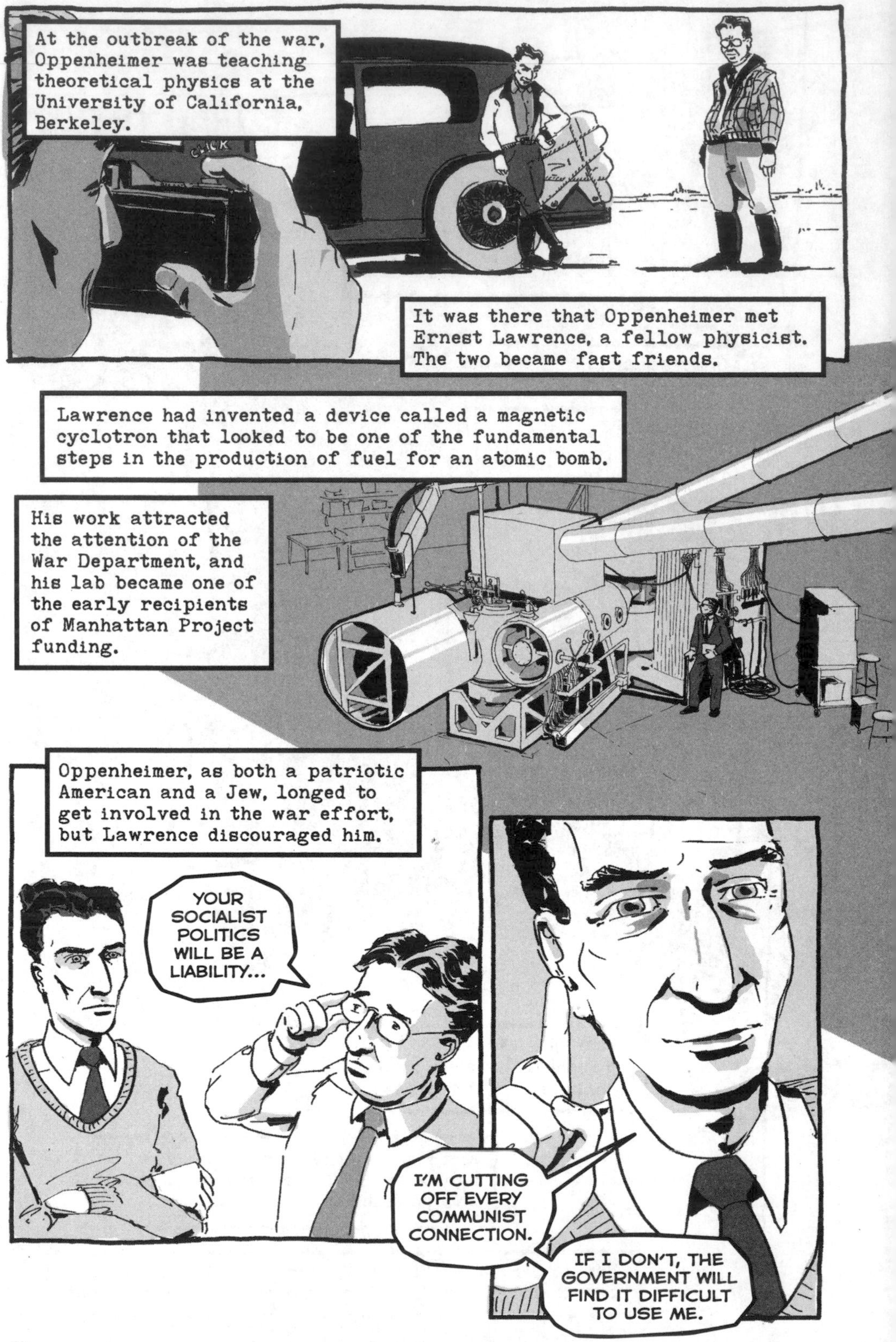
At the outbreak of the war, Oppenheimer was teaching theoretical physics at the University of California, Berkeley.
CLICK
It was there that Oppenheimer met Ernest Lawrence, a fellow physicist. The two became fast friends.
Lawrence had invented a device called a magnetic cyclotron that looked to be one of the fundamental steps in the production of fuel for an atomic bomb.
His work attracted the attention of the War Department, and his lab became one of the early recipients of Manhattan Project funding.
Oppenheimer, as both a patriotic American and a Jew, longed to get involved in the war effort, but Lawrence discouraged him.
YOUR SOCIALIST POLITICS WILL BE A LIABILITY...
I'M CUTTING OFF EVERY COMMUNIST CONNECTION.
IF I DON'T, THE GOVERNMENT WILL FIND IT DIFFICULT TO USE ME.

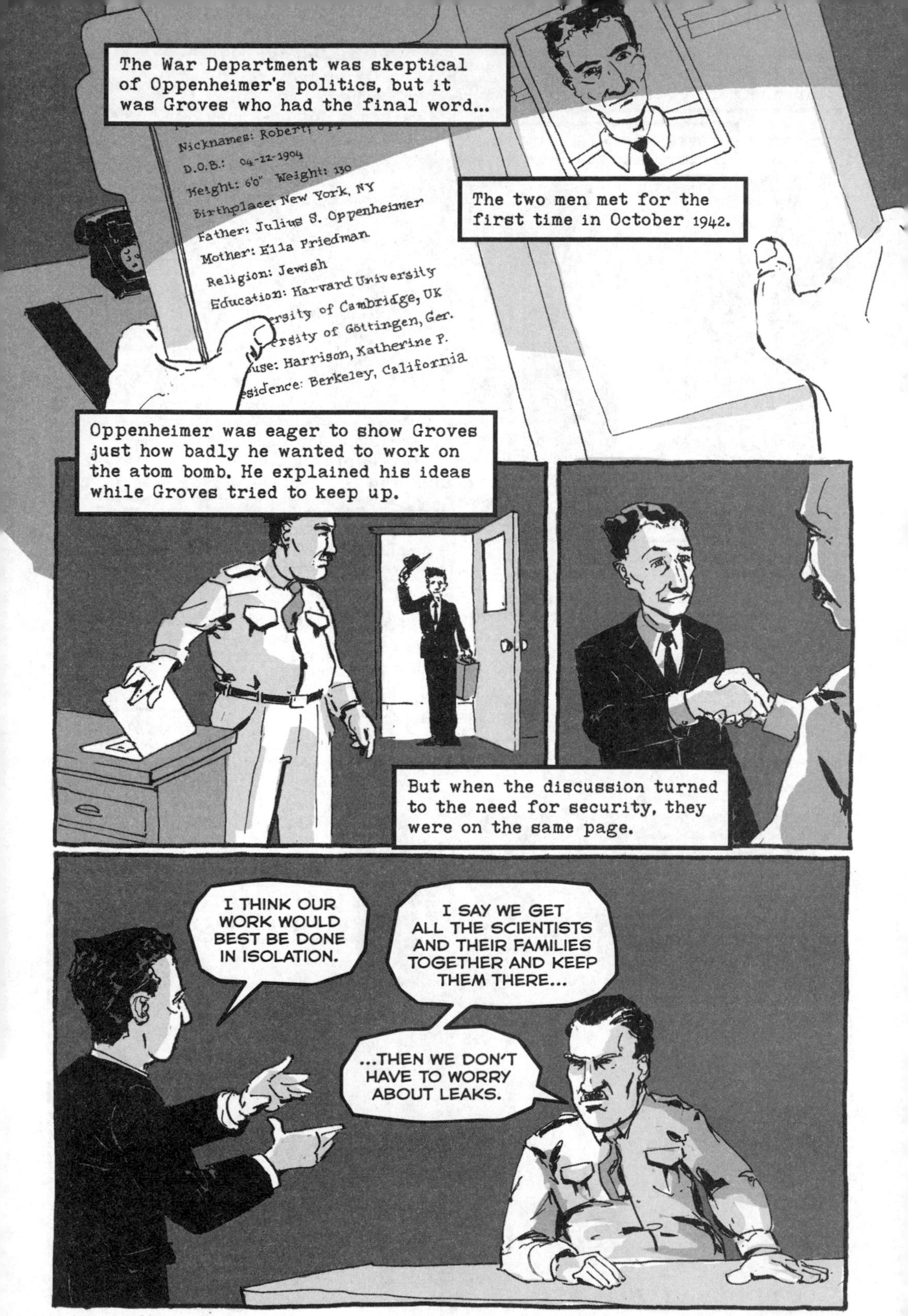

The War Department was skeptical of Oppenheimer's politics, but it was Groves who had the final word...
D.O.B.: 04-22-1904
Height: 6'0" Weight: 130
Birthplace: New York, NY
Father: Julius S. Oppenheimer
Mother: Ella Friedman
Religion: Jewish
Education: Harvard University
of Cambridge, UK
of Göttingen, Ger.
Harrison, Katherine P.
Berkeley, California
The two men met for the first time in October 1942.
Oppenheimer was eager to show Groves just how badly he wanted to work on the atom bomb. He explained his ideas while Groves tried to keep up.
But when the discussion turned to the need for security, they were on the same page.
I THINK OUR WORK WOULD BEST BE DONE IN ISOLATION.
I SAY WE GET ALL THE SCIENTISTS AND THEIR FAMILIES TOGETHER AND KEEP THEM THERE...
...THEN WE DON'T HAVE TO WORRY ABOUT LEAKS.

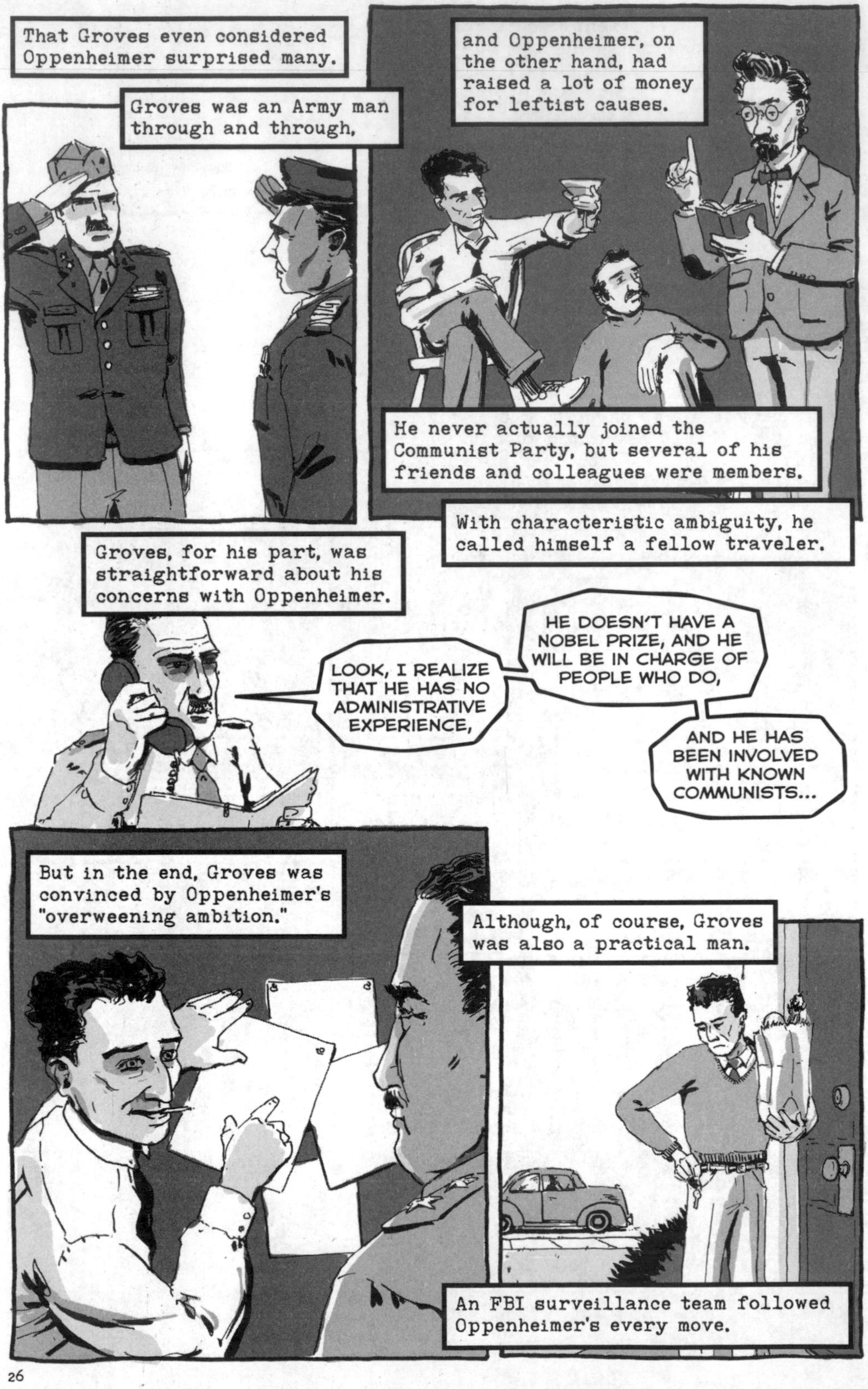
That Groves even considered Oppenheimer surprised many.
Groves was an Army man through and through,
and Oppenheimer, on the other hand, had raised a lot of money for leftist causes.
He never actually joined the Communist Party, but several of his friends and colleagues were members.
With characteristic ambiguity, he called himself a fellow traveler.
Groves, for his part, was straightforward about his concerns with Oppenheimer.
LOOK, I REALIZE THAT HE HAS NO ADMINISTRATIVE EXPERIENCE,
HE DOESN'T HAVE A NOBEL PRIZE, AND HE WILL BE IN CHARGE OF PEOPLE WHO DO,
AND HE HAS BEEN INVOLVED WITH KNOWN COMMUNISTS...
But in the end, Groves was convinced by Oppenheimer's "overweening ambition."
Although, of course, Groves was also a practical man.
An FBI surveillance team followed Oppenheimer's every move.

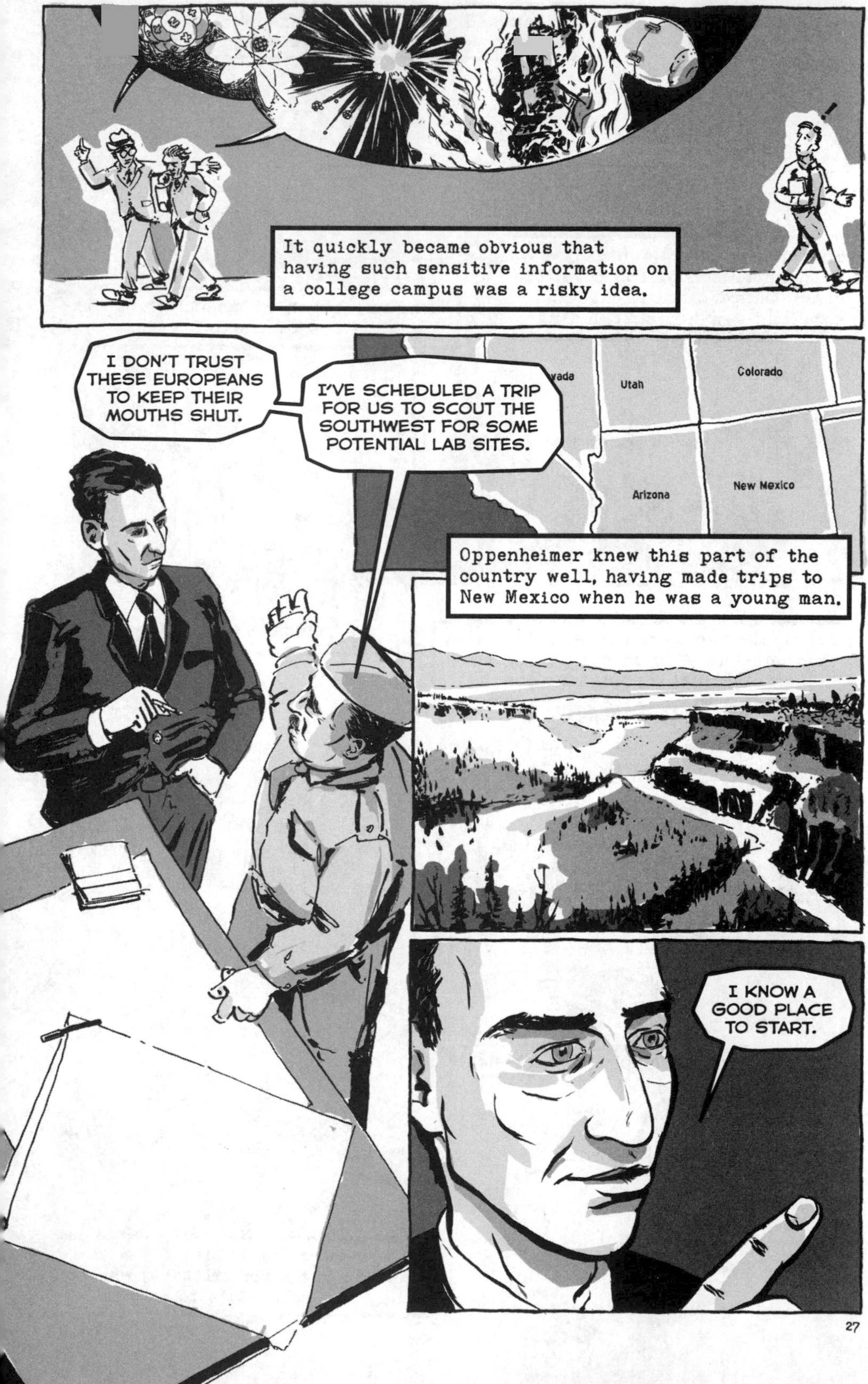
It quickly became obvious that having such sensitive information on a college campus was a risky idea.
I DON'T TRUST THESE EUROPEANS TO KEEP THEIR MOUTHS SHUT.
I'VE SCHEDULED A TRIP FOR US TO SCOUT THE SOUTHWEST FOR SOME POTENTIAL LAB SITES.
Utah
Colorado
Arizona
New Mexico
Oppenheimer knew this part of the country well, having made trips to New Mexico when he was a young man.
I KNOW A GOOD PLACE TO START.

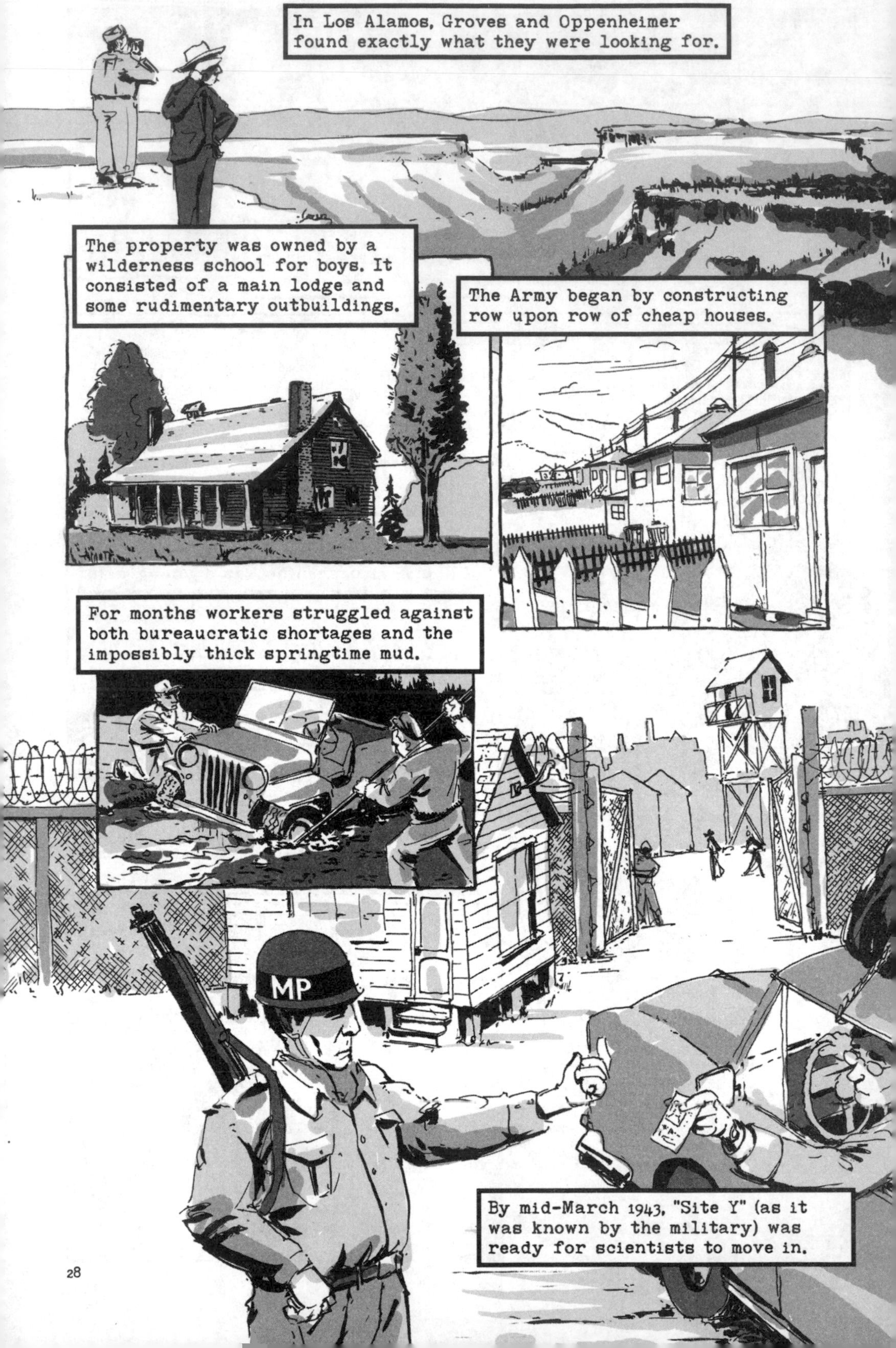
In Los Alamos, Groves and Oppenheimer found exactly what they were looking for.
The property was owned by a wilderness school for boys. It consisted of a main lodge and some rudimentary outbuildings.
The Army began by constructing row upon row of cheap houses.
For months workers struggled against both bureaucratic shortages and the impossibly thick springtime mud.
MP
By mid-March 1943, "Site Y" (as it was known by the military) was ready for scientists to move in.

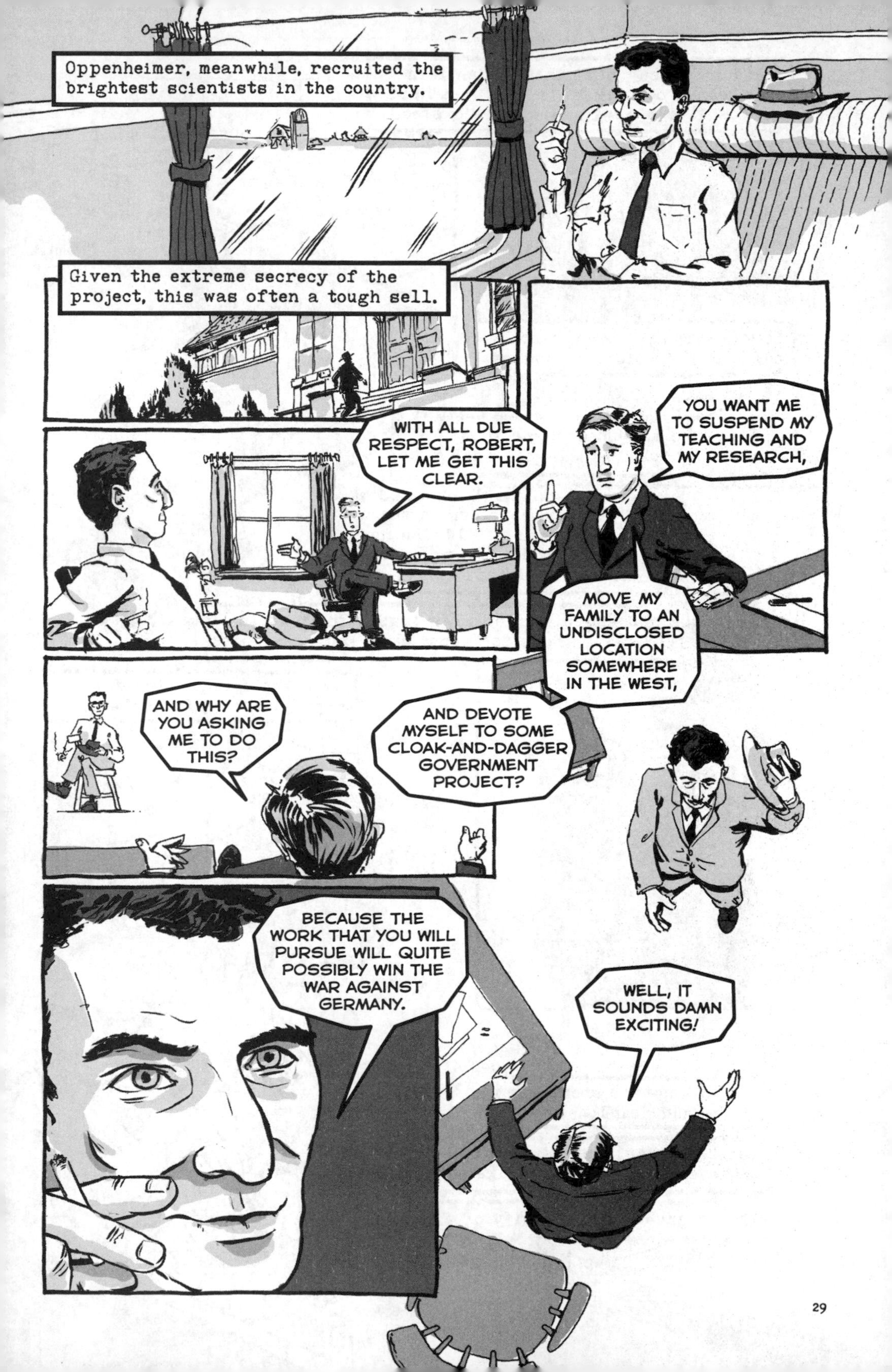
Oppenheimer, meanwhile, recruited the brightest scientists in the country.
Given the extreme secrecy of the project, this was often a tough sell.
WITH ALL DUE RESPECT, ROBERT, LET ME GET THIS CLEAR.
YOU WANT ME TO SUSPEND MY TEACHING AND MY RESEARCH,
MOVE MY FAMILY TO AN UNDISCLOSED LOCATION SOMEWHERE IN THE WEST,
AND DEVOTE MYSELF TO SOME CLOAK-AND-DAGGER GOVERNMENT PROJECT?
AND WHY ARE YOU ASKING ME TO DO THIS?
BECAUSE THE WORK THAT YOU WILL PURSUE WILL QUITE POSSIBLY WIN THE WAR AGAINST GERMANY.
WELL, IT SOUNDS DAMN EXCITING!

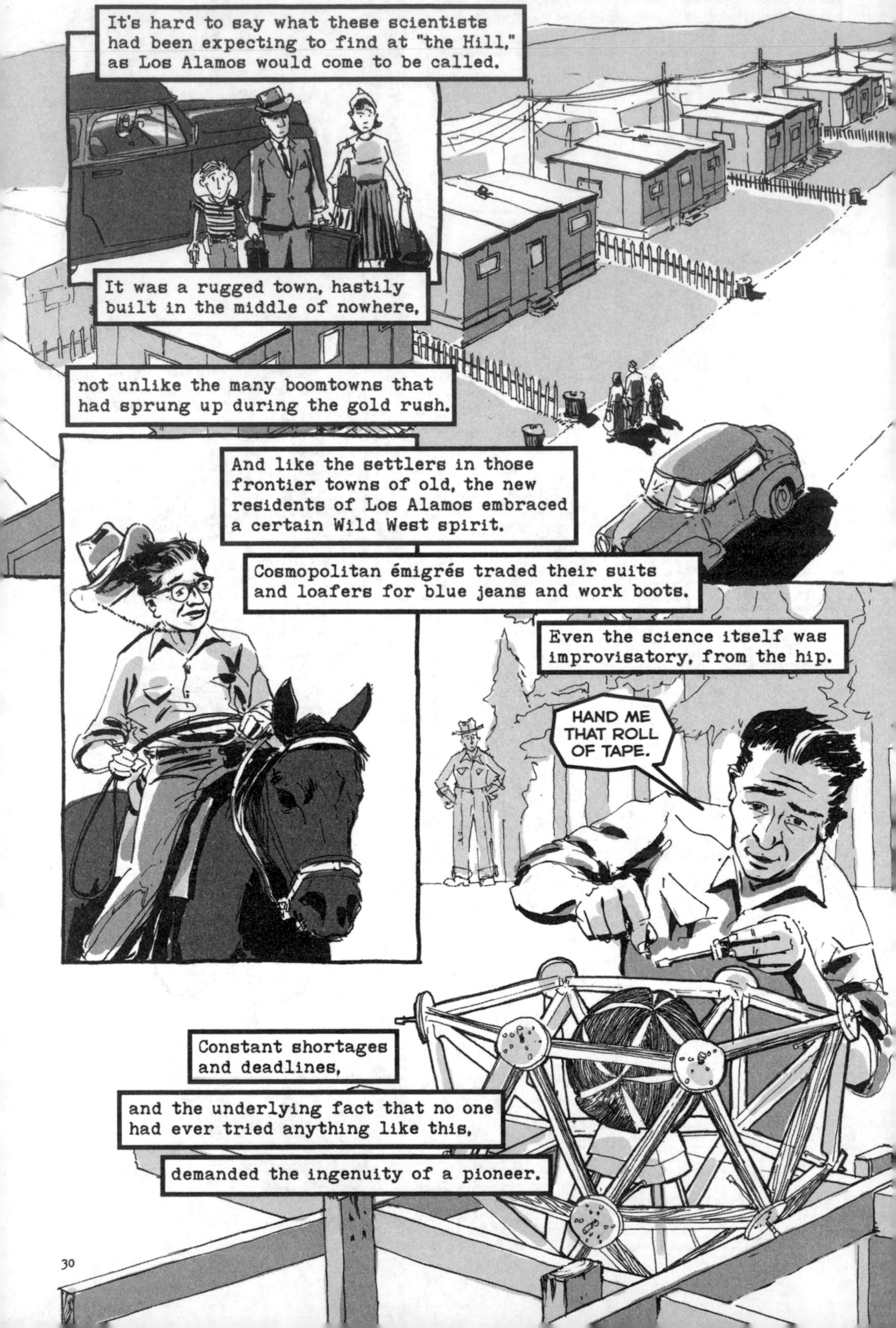

It's hard to say what these scientists had been expecting to find at "the Hill," as Los Alamos would come to be called.
It was a rugged town, hastily built in the middle of nowhere,
not unlike the many boomtowns that had sprung up during the gold rush.
And like the settlers in those frontier towns of old, the new residents of Los Alamos embraced a certain Wild West spirit.
Cosmopolitan émigrés traded their suits and loafers for blue jeans and work boots.
Even the science itself was improvisatory, from the hip.
HAND ME THAT ROLL OF TAPE.
Constant shortages and deadlines,
and the underlying fact that no one had ever tried anything like this,
demanded the ingenuity of a pioneer.

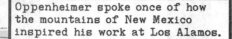

Oppenheimer spoke once of how the mountains of New Mexico inspired his work at Los Alamos.

Perhaps he was speaking of the beautiful vistas and fresh alpine air.

Or maybe he sensed that the harsh, rocky vastness of a place like the American West was the only appropriate birthing ground for a force as elemental as an atomic bomb.

Either way, the Manhattan Project echoes an enduring American mythos:

that with enough money, hard work, open space, and inventiveness, anything is possible.

Such breathless progress would leave little time to consider the consequences.

There was work to be done.

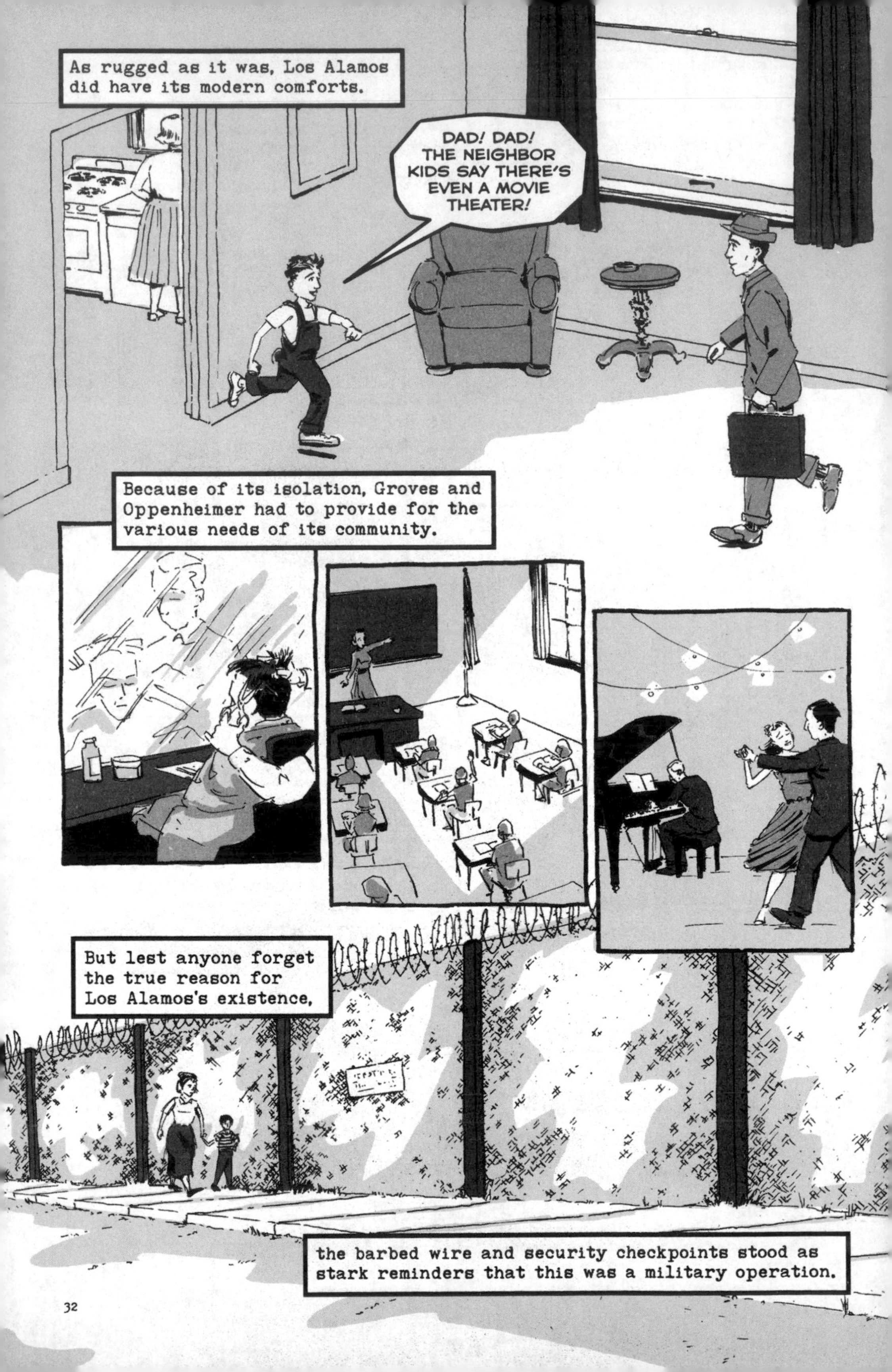
As rugged as it was, Los Alamos did have its modern comforts.
DAD! DAD! THE NEIGHBOR KIDS SAY THERE'S EVEN A MOVIE THEATER!
Because of its isolation, Groves and Oppenheimer had to provide for the various needs of its community.
But lest anyone forget the true reason for Los Alamos's existence,
the barbed wire and security checkpoints stood as stark reminders that this was a military operation.

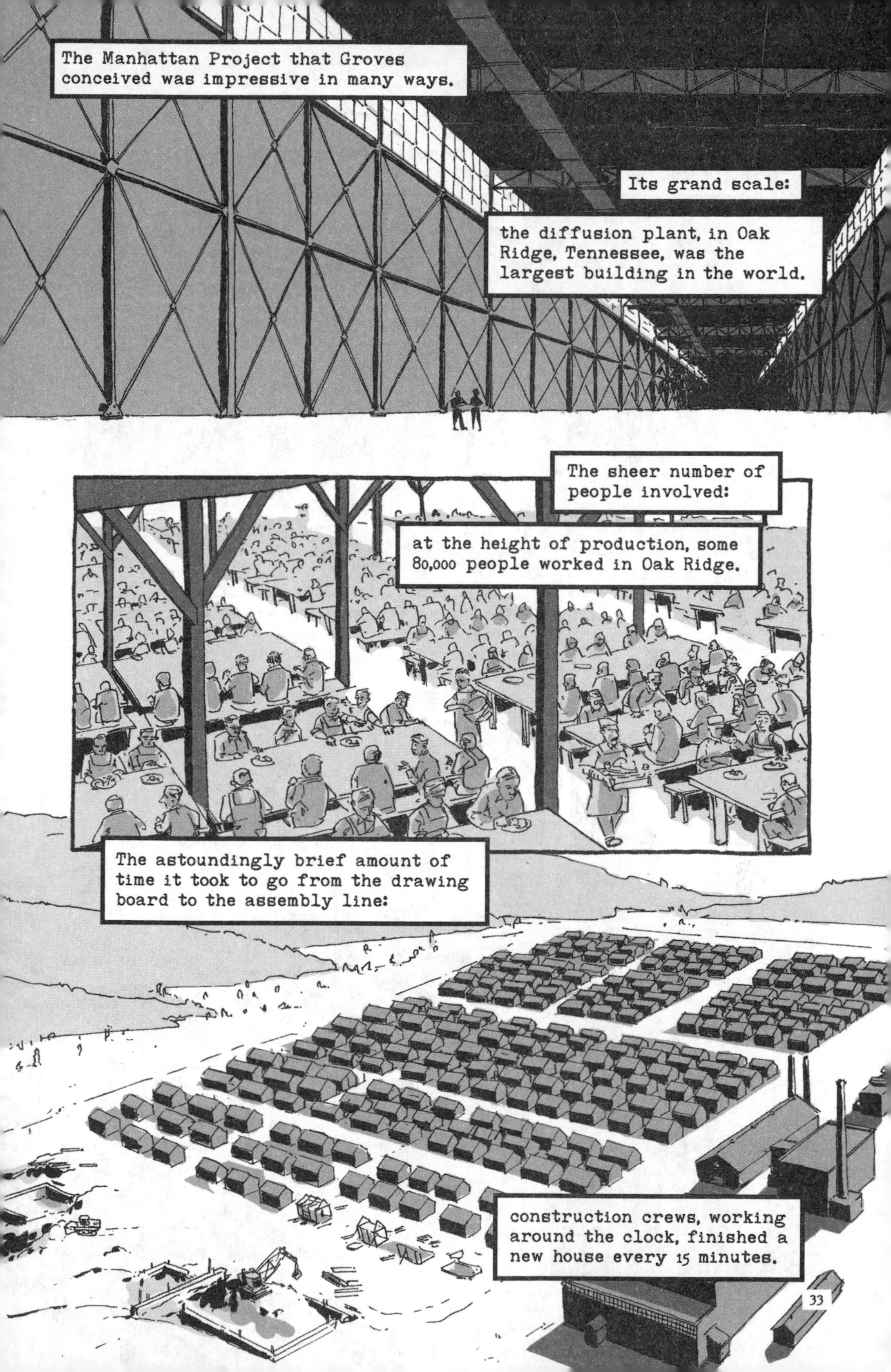

The Manhattan Project that Groves conceived was impressive in many ways.
Its grand scale:
the diffusion plant, in Oak Ridge, Tennessee, was the largest building in the world.
The sheer number of people involved:
at the height of production, some 80,000 people worked in Oak Ridge.
The astoundingly brief amount of time it took to go from the drawing board to the assembly line:
construction crews, working around the clock, finished a new house every 15 minutes.

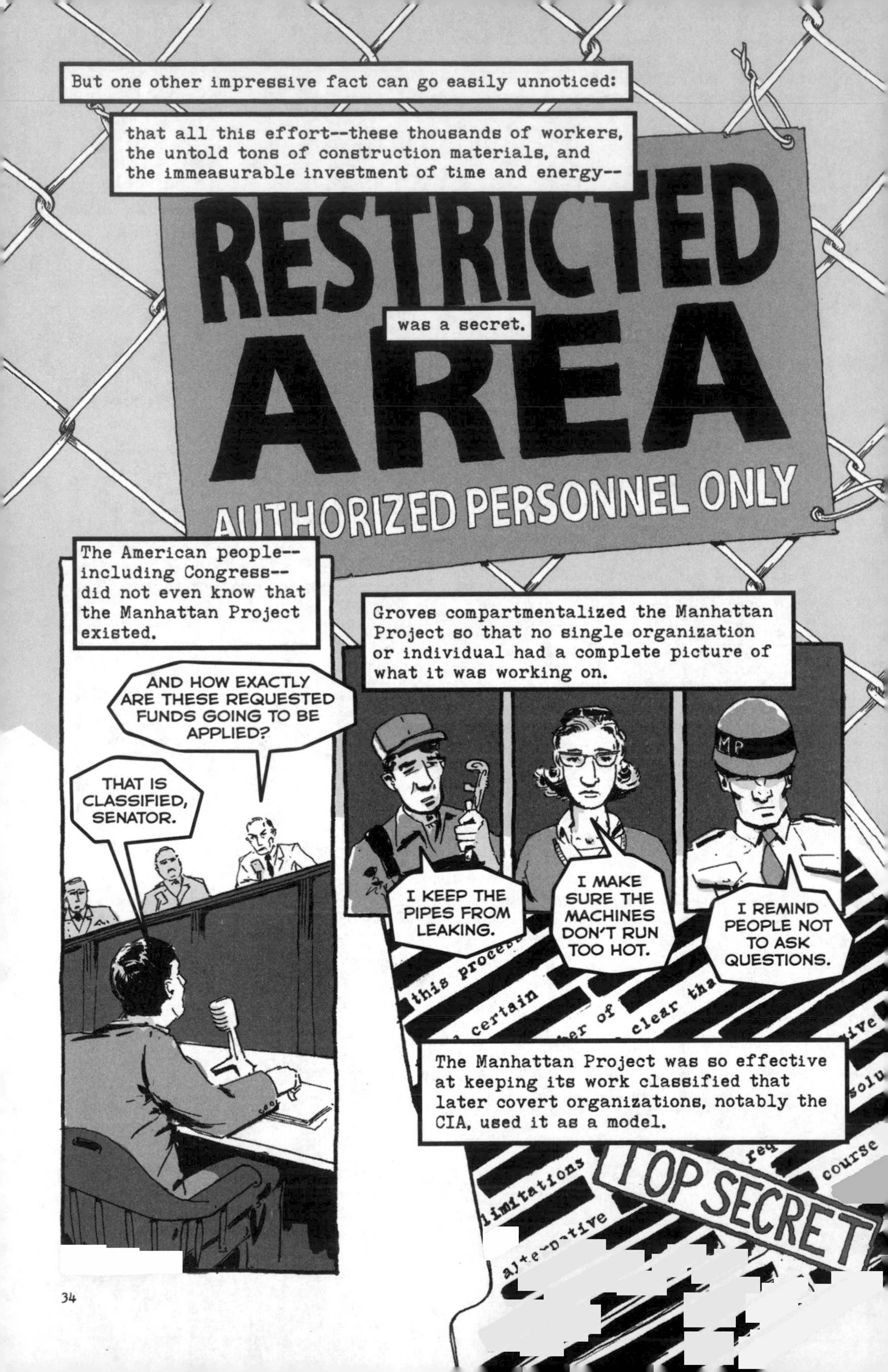
But one other impressive fact can go easily unnoticed:
that all this effort--these thousands of workers, the untold tons of construction materials, and the immeasurable investment of time and energy--
RESTRICTED
was a secret.
AREA
AUTHORIZED PERSONNEL ONLY
The American people--including Congress--did not even know that the Manhattan Project existed.
AND HOW EXACTLY ARE THESE REQUESTED FUNDS GOING TO BE APPLIED?
THAT IS CLASSIFIED, SENATOR.
Groves compartmentalized the Manhattan Project so that no single organization or individual had a complete picture of what it was working on.
I KEEP THE PIPES FROM LEAKING.
I MAKE SURE THE MACHINES DON'T RUN TOO HOT.
I REMIND PEOPLE NOT TO ASK QUESTIONS.
certain
The Manhattan Project was so effective at keeping its work classified that later covert organizations, notably the CIA, used it as a model.
limitations
course
TOP SECRET

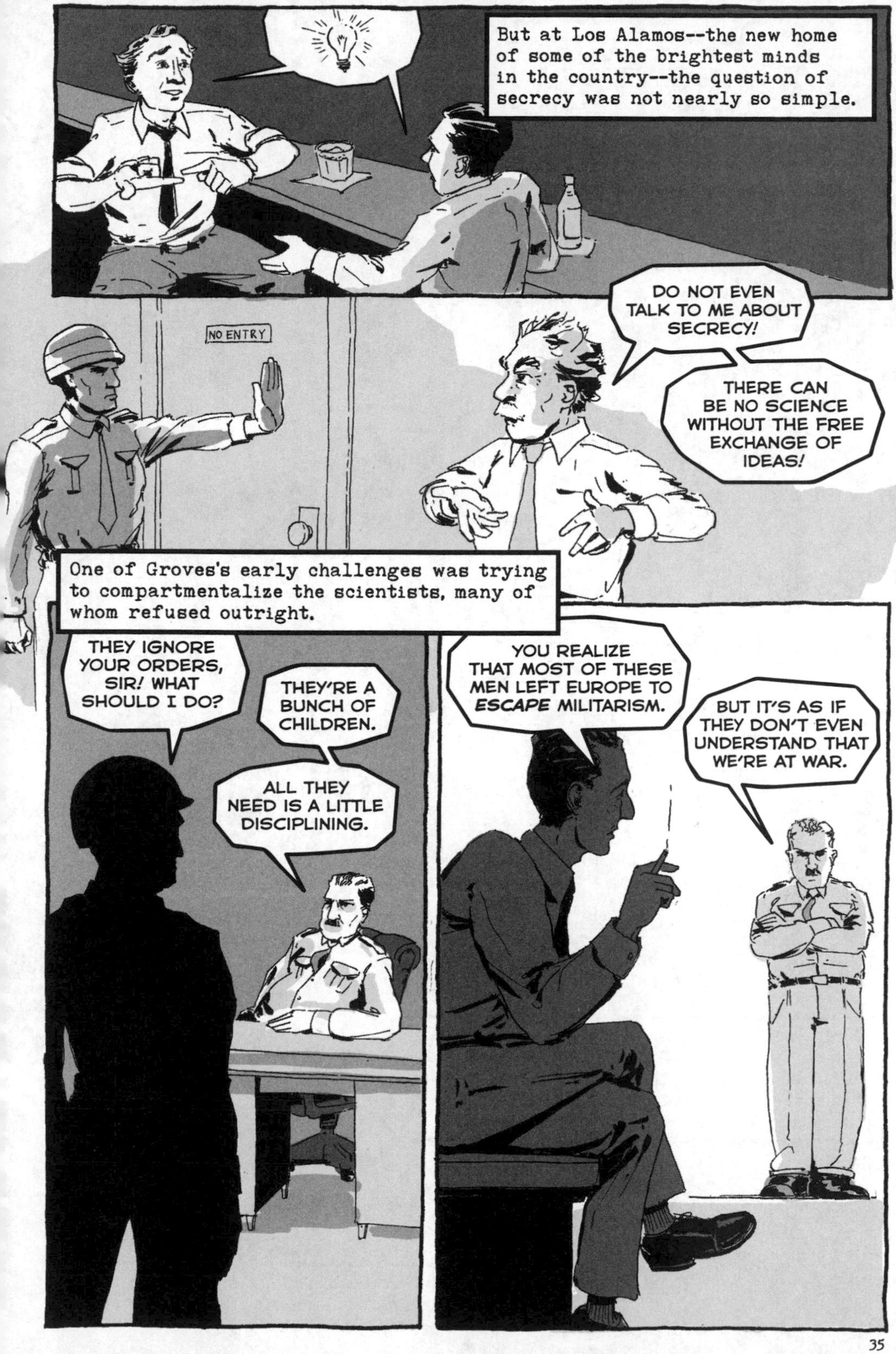
But at Los Alamos--the new home of some of the brightest minds in the country--the question of secrecy was not nearly so simple.
NO ENTRY
DO NOT EVEN TALK TO ME ABOUT SECRECY!
THERE CAN BE NO SCIENCE WITHOUT THE FREE EXCHANGE OF IDEAS!
One of Groves's early challenges was trying to compartmentalize the scientists, many of whom refused outright.
THEY IGNORE YOUR ORDERS, SIR! WHAT SHOULD I DO?
THEY'RE A BUNCH OF CHILDREN.
ALL THEY NEED IS A LITTLE DISCIPLINING.
YOU REALIZE THAT MOST OF THESE MEN LEFT EUROPE TO *ESCAPE* MILITARISM.
BUT IT'S AS IF THEY DON'T EVEN UNDERSTAND THAT WE'RE AT WAR.

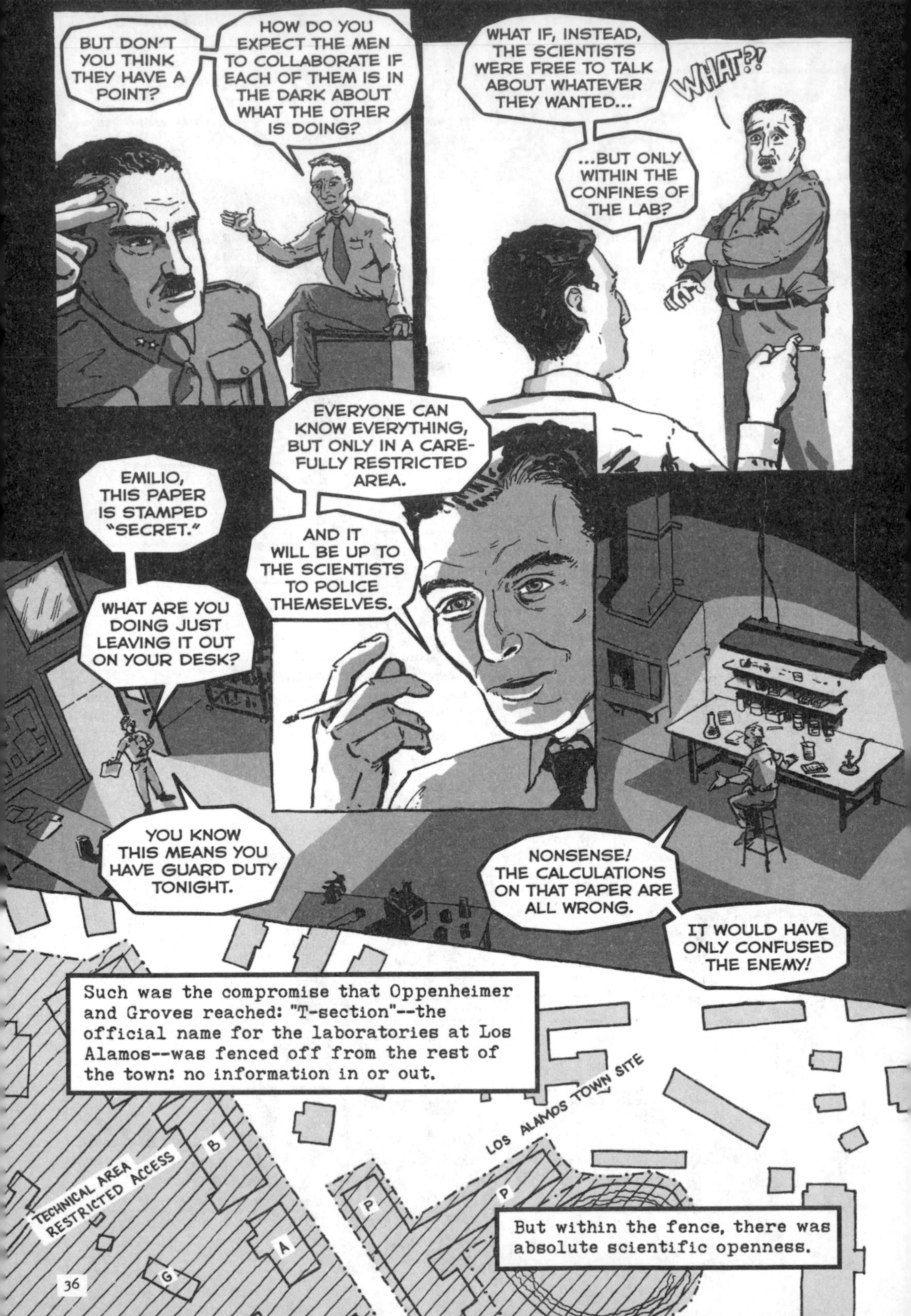

BUT DON'T YOU THINK THEY HAVE A POINT?
HOW DO YOU EXPECT THE MEN TO COLLABORATE IF EACH OF THEM IS IN THE DARK ABOUT WHAT THE OTHER IS DOING?
WHAT IF, INSTEAD, THE SCIENTISTS WERE FREE TO TALK ABOUT WHATEVER THEY WANTED...
WHAT?!
...BUT ONLY WITHIN THE CONFINES OF THE LAB?
EVERYONE CAN KNOW EVERYTHING, BUT ONLY IN A CAREFULLY RESTRICTED AREA.
AND IT WILL BE UP TO THE SCIENTISTS TO POLICE THEMSELVES.
EMILIO, THIS PAPER IS STAMPED "SECRET."
WHAT ARE YOU DOING JUST LEAVING IT OUT ON YOUR DESK?
YOU KNOW THIS MEANS YOU HAVE GUARD DUTY TONIGHT.
NONSENSE! THE CALCULATIONS ON THAT PAPER ARE ALL WRONG.
IT WOULD HAVE ONLY CONFUSED THE ENEMY!
Such was the compromise that Oppenheimer and Groves reached: "T-section"--the official name for the laboratories at Los Alamos--was fenced off from the rest of the town: no information in or out.
LOS ALAMOS TOWN SITE
TECHNICAL AREA RESTRICTED ACCESS
B
P
P
A
G
But within the fence, there was absolute scientific openness.

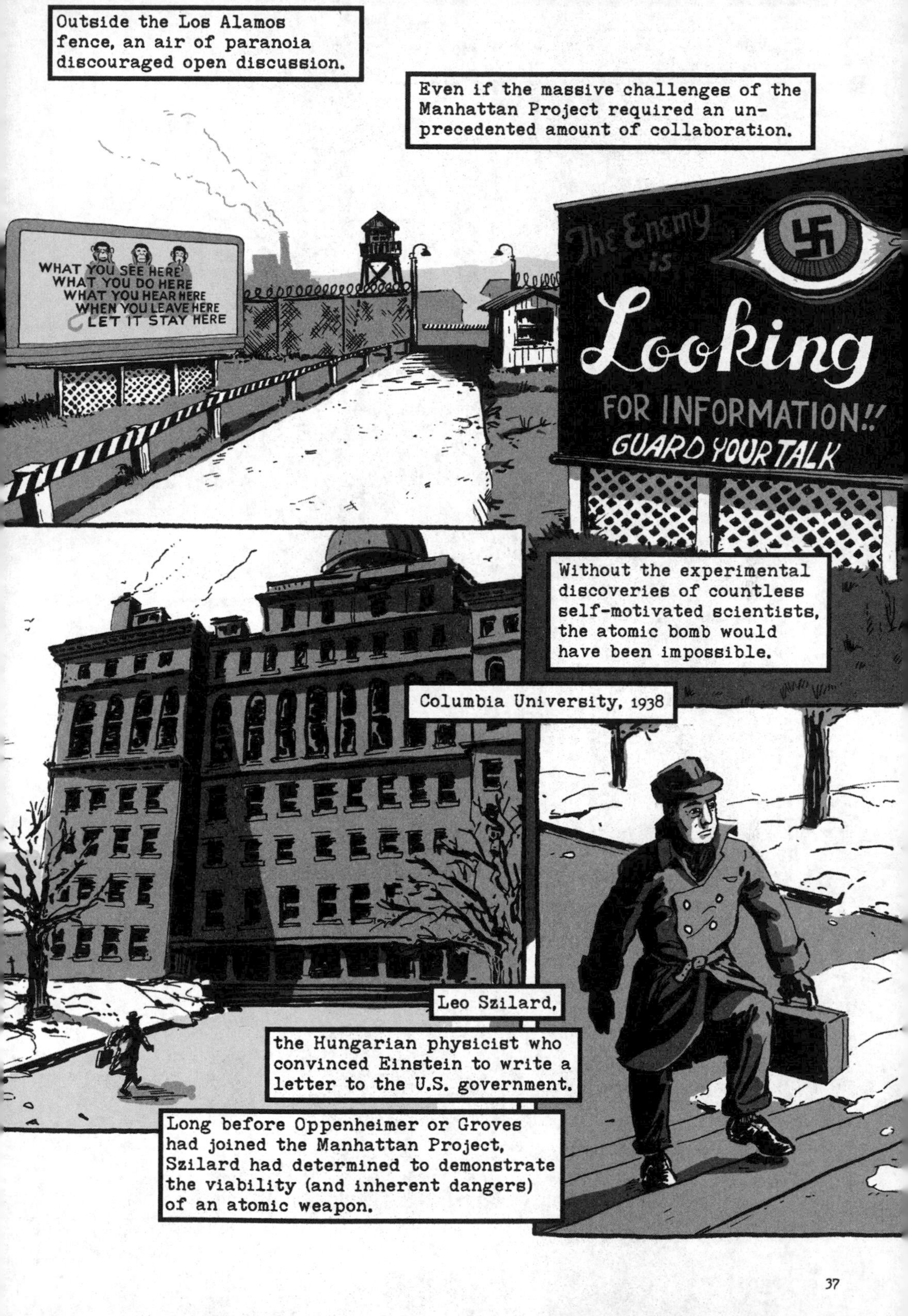
Outside the Los Alamos fence, an air of paranoia discouraged open discussion.
Even if the massive challenges of the Manhattan Project required an unprecedented amount of collaboration.
WHAT YOU SEE HERE
WHAT YOU DO HERE
WHAT YOU HEAR HERE
WHEN YOU LEAVE HERE
LET IT STAY HERE
The Enemy is Looking FOR INFORMATION!!
GUARD YOUR TALK
Without the experimental discoveries of countless self-motivated scientists, the atomic bomb would have been impossible.
Columbia University, 1938
Leo Szilard,
the Hungarian physicist who convinced Einstein to write a letter to the U.S. government.
Long before Oppenheimer or Groves had joined the Manhattan Project, Szilard had determined to demonstrate the viability (and inherent dangers) of an atomic weapon.

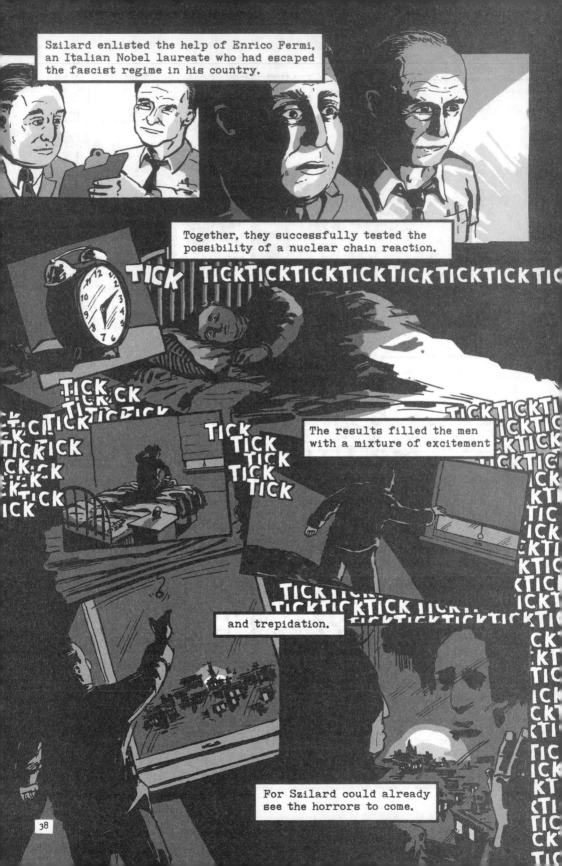

Szilard enlisted the help of Enrico Fermi, an Italian Nobel laureate who had escaped the fascist regime in his country.

Together, they successfully tested the possibility of a nuclear chain reaction.

TICK TICKTICKTICKTICKTICKTICKTICKTIC

TICK
TICK CK
TICK ICK
TICKICK
CK CK
CKTICK
ICK

TICK
TICK
TICK
TICK

The results filled the men with a mixture of excitement

TICKTICKTICK TICKTICKTICKTICKTICKTICK

and trepidation.

For Szilard could already see the horrors to come.

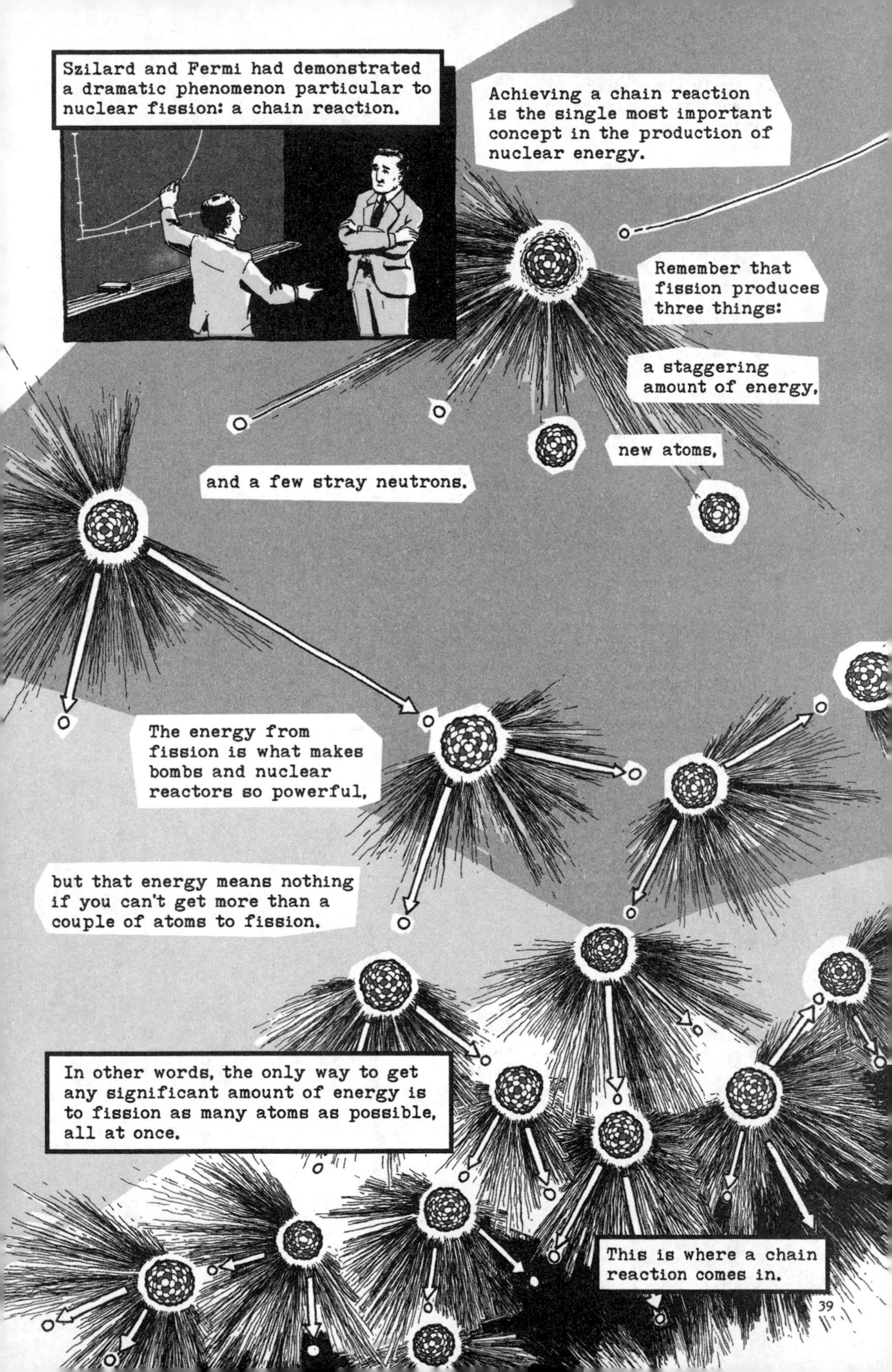

Szilard and Fermi had demonstrated a dramatic phenomenon particular to nuclear fission: a chain reaction.
Achieving a chain reaction is the single most important concept in the production of nuclear energy.
Remember that fission produces three things:
a staggering amount of energy,
new atoms,
and a few stray neutrons.
The energy from fission is what makes bombs and nuclear reactors so powerful,
but that energy means nothing if you can't get more than a couple of atoms to fission.
In other words, the only way to get any significant amount of energy is to fission as many atoms as possible, all at once.
This is where a chain reaction comes in.

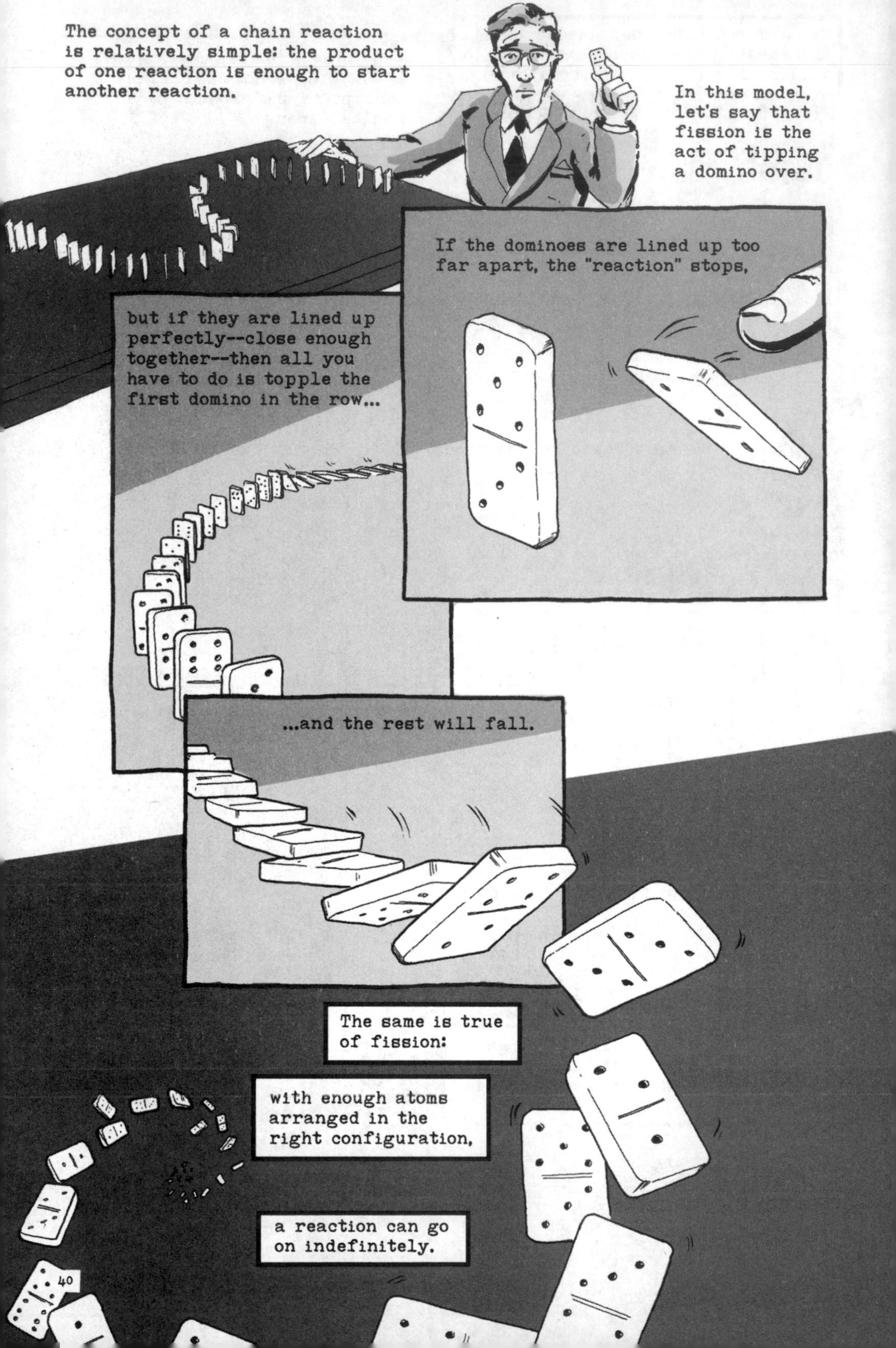
The concept of a chain reaction is relatively simple: the product of one reaction is enough to start another reaction.
In this model, let's say that fission is the act of tipping a domino over.
If the dominoes are lined up too far apart, the "reaction" stops,
but if they are lined up perfectly--close enough together--then all you have to do is topple the first domino in the row...
...and the rest will fall.
The same is true of fission:
with enough atoms arranged in the right configuration,
a reaction can go on indefinitely.

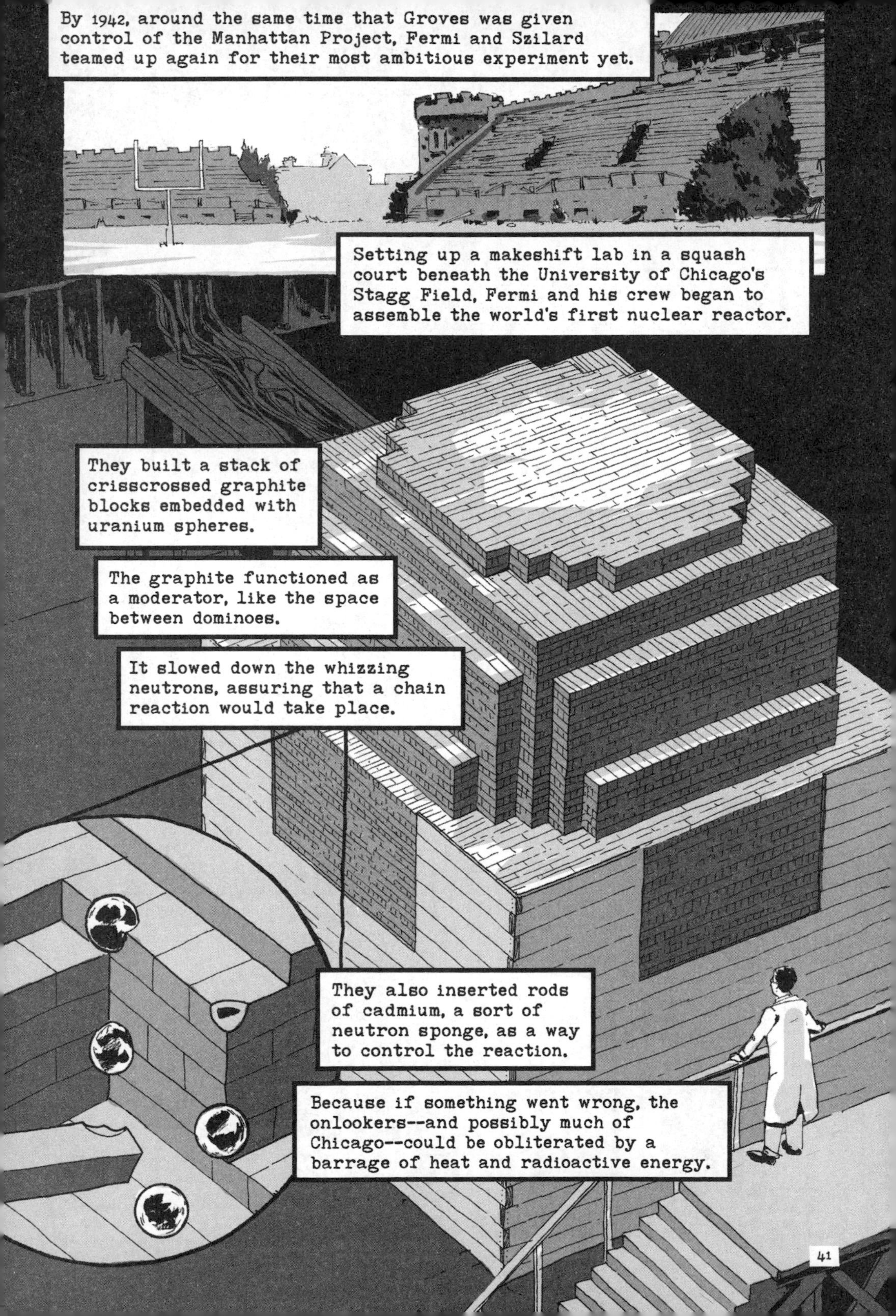
By 1942, around the same time that Groves was given control of the Manhattan Project, Fermi and Szilard teamed up again for their most ambitious experiment yet.
Setting up a makeshift lab in a squash court beneath the University of Chicago's Stagg Field, Fermi and his crew began to assemble the world's first nuclear reactor.
They built a stack of crisscrossed graphite blocks embedded with uranium spheres.
The graphite functioned as a moderator, like the space between dominoes.
It slowed down the whizzing neutrons, assuring that a chain reaction would take place.
They also inserted rods of cadmium, a sort of neutron sponge, as a way to control the reaction.
Because if something went wrong, the onlookers--and possibly much of Chicago--could be obliterated by a barrage of heat and radioactive energy.

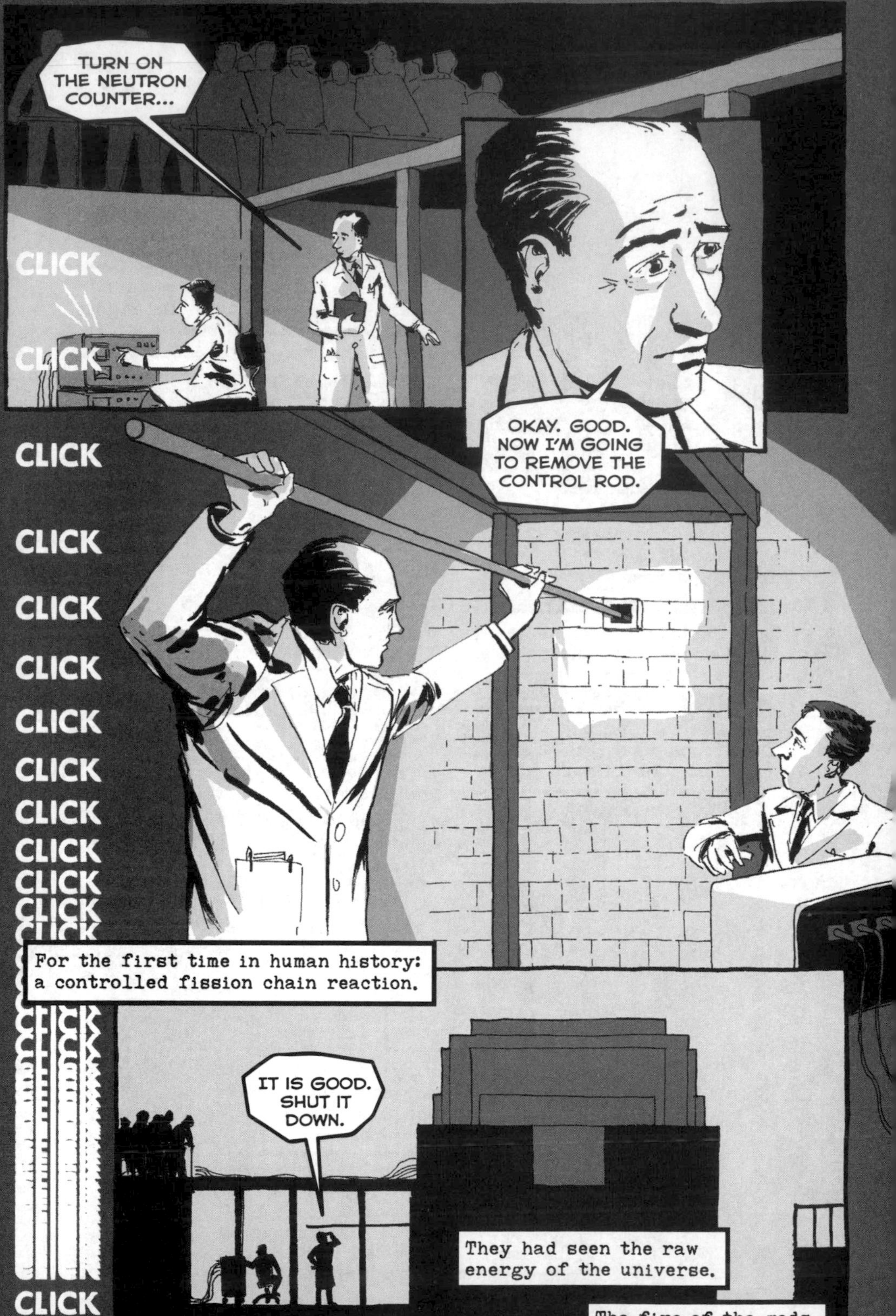
TURN ON THE NEUTRON COUNTER...
CLICK
CLICK
OKAY. GOOD. NOW I'M GOING TO REMOVE THE CONTROL ROD.
CLICK
CLICK
CLICK
CLICK
CLICK
CLICK
CLICK
CLICK
CLICK
For the first time in human history: a controlled fission chain reaction.
IT IS GOOD. SHUT IT DOWN.
CLICK
They had seen the raw energy of the universe.
The fire of the gods.

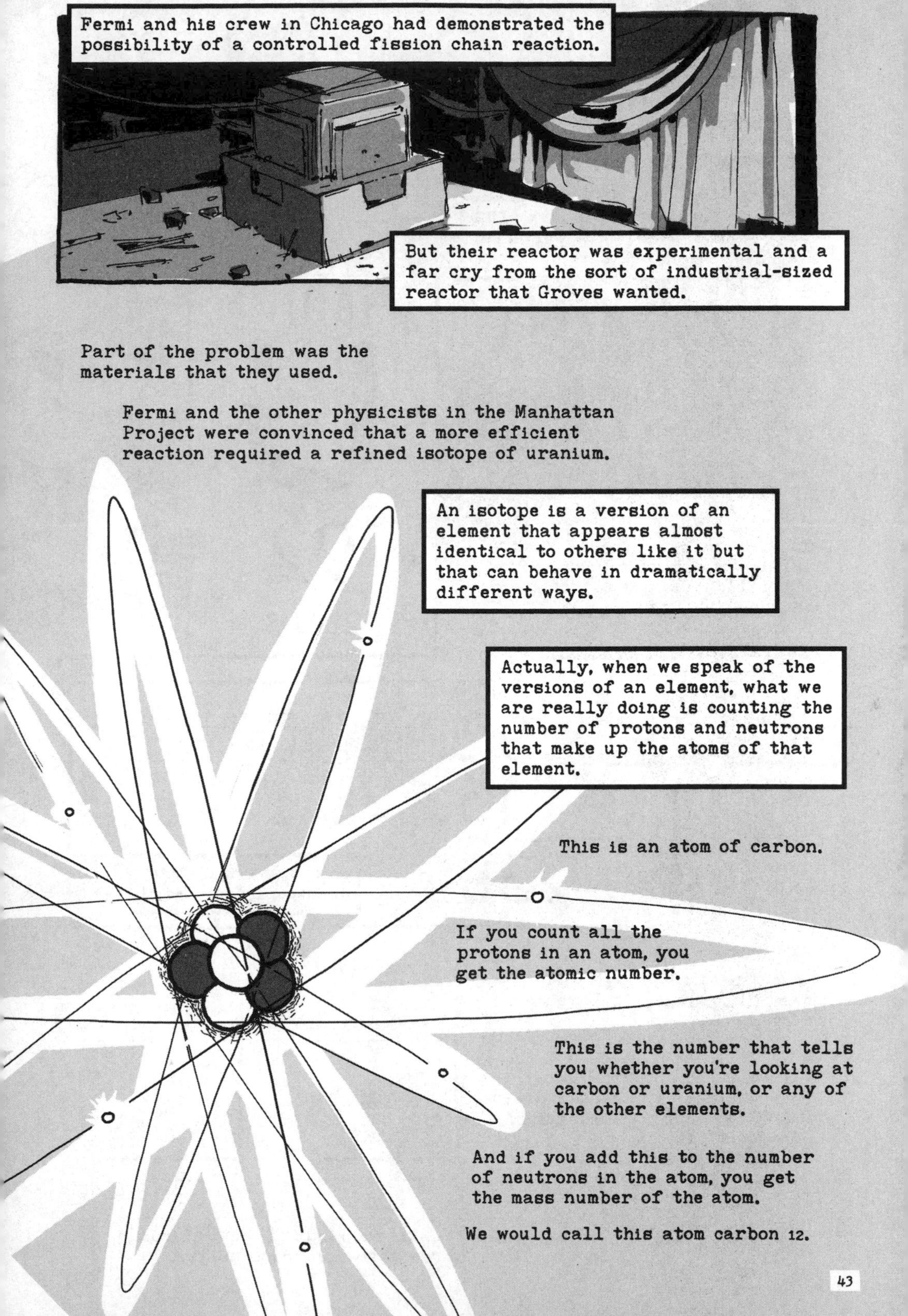
Fermi and his crew in Chicago had demonstrated the possibility of a controlled fission chain reaction.
But their reactor was experimental and a far cry from the sort of industrial-sized reactor that Groves wanted.
Part of the problem was the materials that they used.
Fermi and the other physicists in the Manhattan Project were convinced that a more efficient reaction required a refined isotope of uranium.
An isotope is a version of an element that appears almost identical to others like it but that can behave in dramatically different ways.
Actually, when we speak of the versions of an element, what we are really doing is counting the number of protons and neutrons that make up the atoms of that element.
This is an atom of carbon.
If you count all the protons in an atom, you get the atomic number.
This is the number that tells you whether you're looking at carbon or uranium, or any of the other elements.
And if you add this to the number of neutrons in the atom, you get the mass number of the atom.
We would call this atom carbon 12.

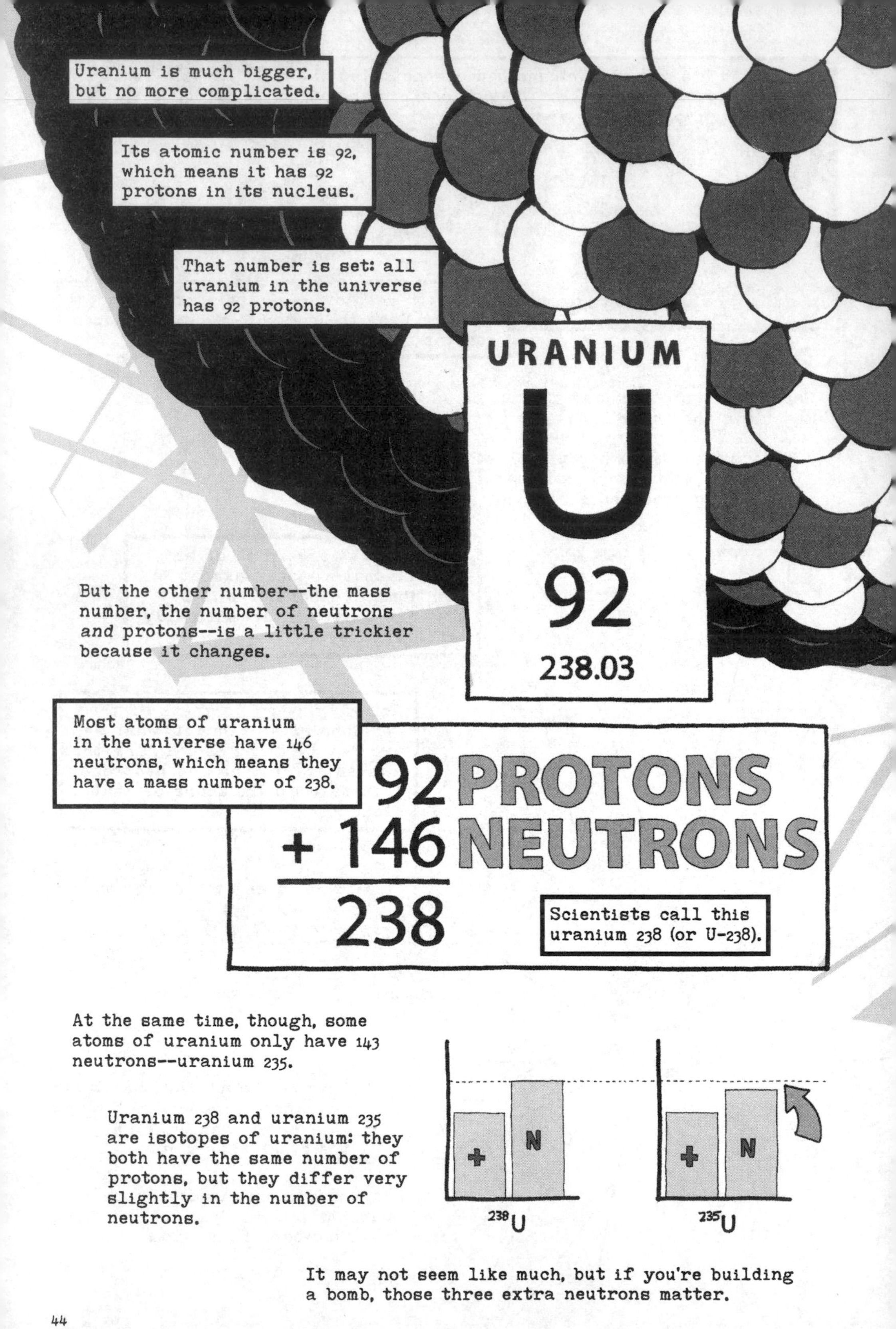

Uranium is much bigger, but no more complicated.

Its atomic number is 92, which means it has 92 protons in its nucleus.

That number is set: all uranium in the universe has 92 protons.

URANIUM

U
92
238.03

But the other number--the mass number, the number of neutrons and protons--is a little trickier because it changes.

Most atoms of uranium in the universe have 146 neutrons, which means they have a mass number of 238.

$$\begin{array}{r} 92 \\ + 146 \\ \hline 238 \end{array}$$ **PROTONS NEUTRONS**

Scientists call this uranium 238 (or U-238).

At the same time, though, some atoms of uranium only have 143 neutrons--uranium 235.

Uranium 238 and uranium 235 are isotopes of uranium: they both have the same number of protons, but they differ very slightly in the number of neutrons.

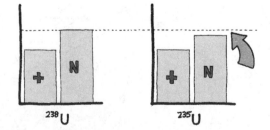

^{238}U ^{235}U

It may not seem like much, but if you're building a bomb, those three extra neutrons matter.

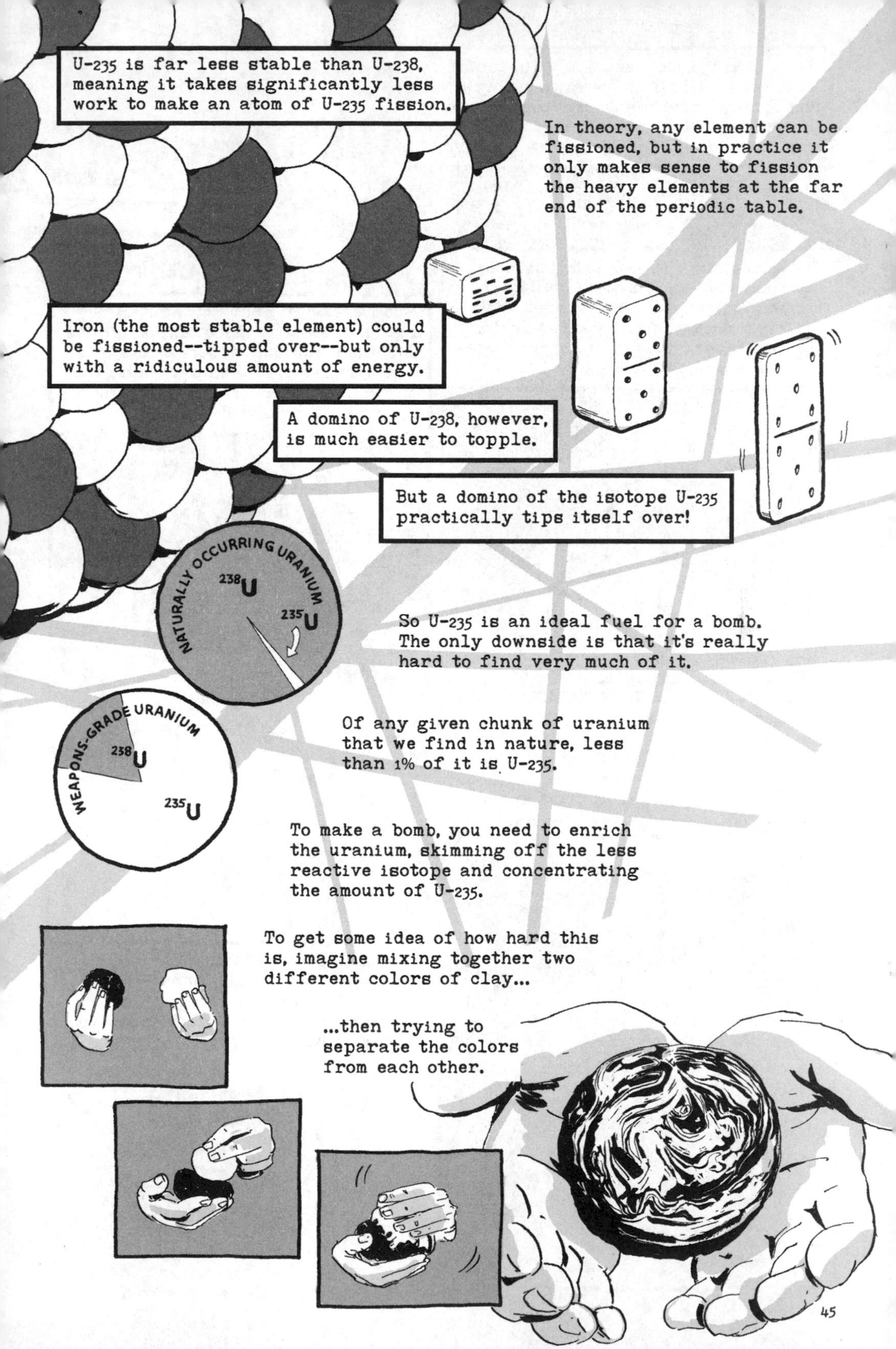
U-235 is far less stable than U-238, meaning it takes significantly less work to make an atom of U-235 fission.
In theory, any element can be fissioned, but in practice it only makes sense to fission the heavy elements at the far end of the periodic table.
Iron (the most stable element) could be fissioned--tipped over--but only with a ridiculous amount of energy.
A domino of U-238, however, is much easier to topple.
But a domino of the isotope U-235 practically tips itself over!
NATURALLY OCCURRING URANIUM
238U
235U
So U-235 is an ideal fuel for a bomb. The only downside is that it's really hard to find very much of it.
WEAPONS-GRADE URANIUM
238U
235U
Of any given chunk of uranium that we find in nature, less than 1% of it is U-235.
To make a bomb, you need to enrich the uranium, skimming off the less reactive isotope and concentrating the amount of U-235.
To get some idea of how hard this is, imagine mixing together two different colors of clay...
...then trying to separate the colors from each other.

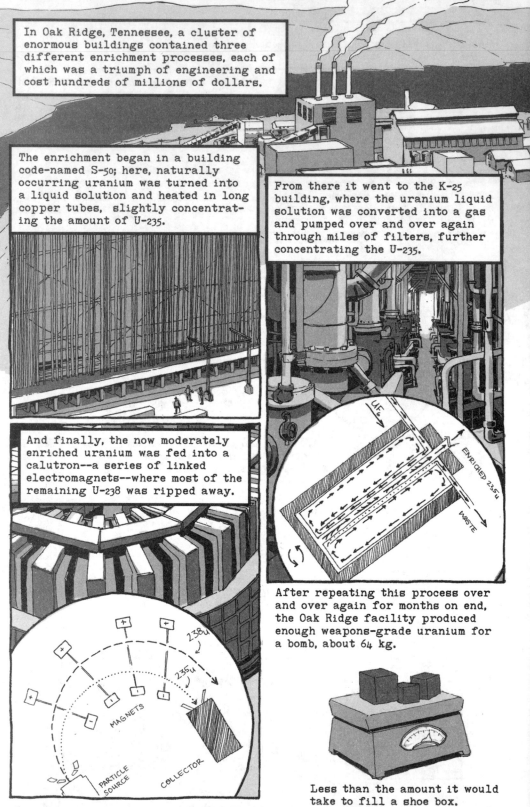

In Oak Ridge, Tennessee, a cluster of enormous buildings contained three different enrichment processes, each of which was a triumph of engineering and cost hundreds of millions of dollars.

The enrichment began in a building code-named S-50; here, naturally occurring uranium was turned into a liquid solution and heated in long copper tubes, slightly concentrating the amount of U-235.

From there it went to the K-25 building, where the uranium liquid solution was converted into a gas and pumped over and over again through miles of filters, further concentrating the U-235.

And finally, the now moderately enriched uranium was fed into a calutron--a series of linked electromagnets--where most of the remaining U-238 was ripped away.

UF₆

ENRICHED 235 u

WASTE

After repeating this process over and over again for months on end, the Oak Ridge facility produced enough weapons-grade uranium for a bomb, about 64 kg.

+ +
+
238u
+
235u
- -
MAGNETS
-
PARTICLE SOURCE   COLLECTOR

Less than the amount it would take to fill a shoe box.

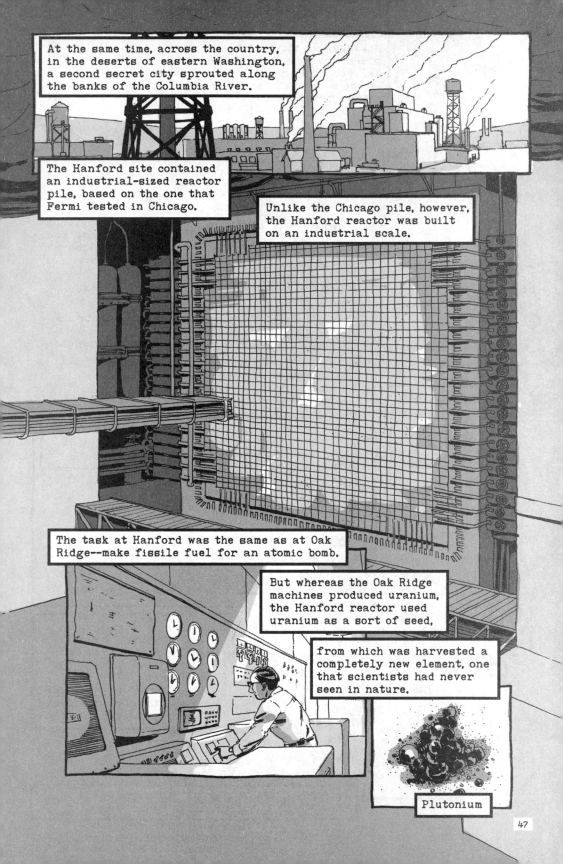

At the same time, across the country, in the deserts of eastern Washington, a second secret city sprouted along the banks of the Columbia River.

The Hanford site contained an industrial-sized reactor pile, based on the one that Fermi tested in Chicago.

Unlike the Chicago pile, however, the Hanford reactor was built on an industrial scale.

The task at Hanford was the same as at Oak Ridge--make fissile fuel for an atomic bomb.

But whereas the Oak Ridge machines produced uranium, the Hanford reactor used uranium as a sort of seed,

from which was harvested a completely new element, one that scientists had never seen in nature.

Plutonium

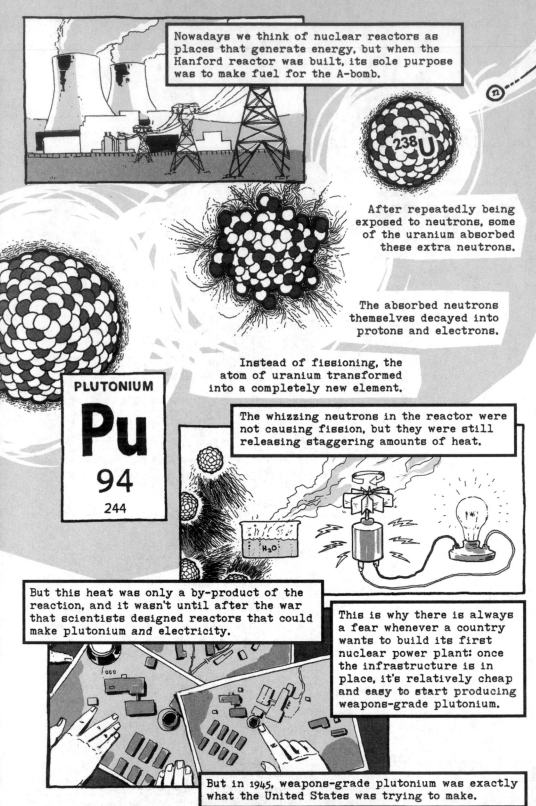

Nowadays we think of nuclear reactors as places that generate energy, but when the Hanford reactor was built, its sole purpose was to make fuel for the A-bomb.

After repeatedly being exposed to neutrons, some of the uranium absorbed these extra neutrons.

The absorbed neutrons themselves decayed into protons and electrons.

Instead of fissioning, the atom of uranium transformed into a completely new element.

PLUTONIUM

Pu

94

244

The whizzing neutrons in the reactor were not causing fission, but they were still releasing staggering amounts of heat.

But this heat was only a by-product of the reaction, and it wasn't until after the war that scientists designed reactors that could make plutonium and electricity.

This is why there is always a fear whenever a country wants to build its first nuclear power plant: once the infrastructure is in place, it's relatively cheap and easy to start producing weapons-grade plutonium.

But in 1945, weapons-grade plutonium was exactly what the United States was trying to make.

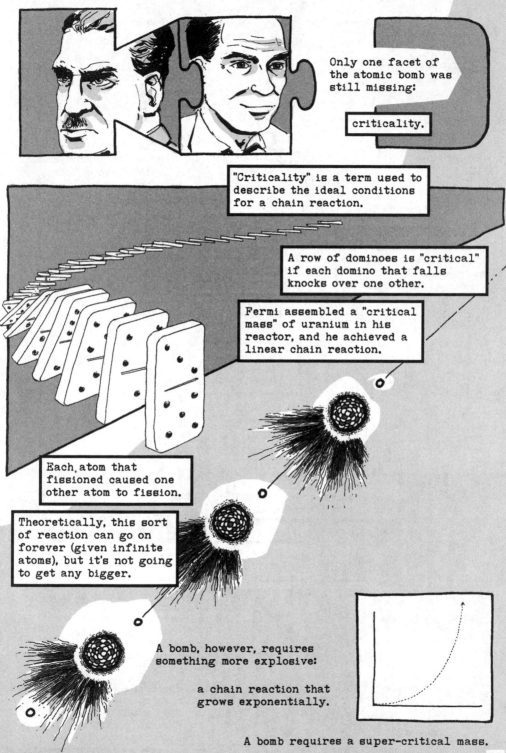

Groves had built the factories necessary to make bomb fuel. Fermi had demonstrated the viability of making a chain reaction.

Only one facet of the atomic bomb was still missing:

criticality.

"Criticality" is a term used to describe the ideal conditions for a chain reaction.

A row of dominoes is "critical" if each domino that falls knocks over one other.

Fermi assembled a "critical mass" of uranium in his reactor, and he achieved a linear chain reaction.

Each atom that fissioned caused one other atom to fission.

Theoretically, this sort of reaction can go on forever (given infinite atoms), but it's not going to get any bigger.

A bomb, however, requires something more explosive:

a chain reaction that grows exponentially.

A bomb requires a super-critical mass.

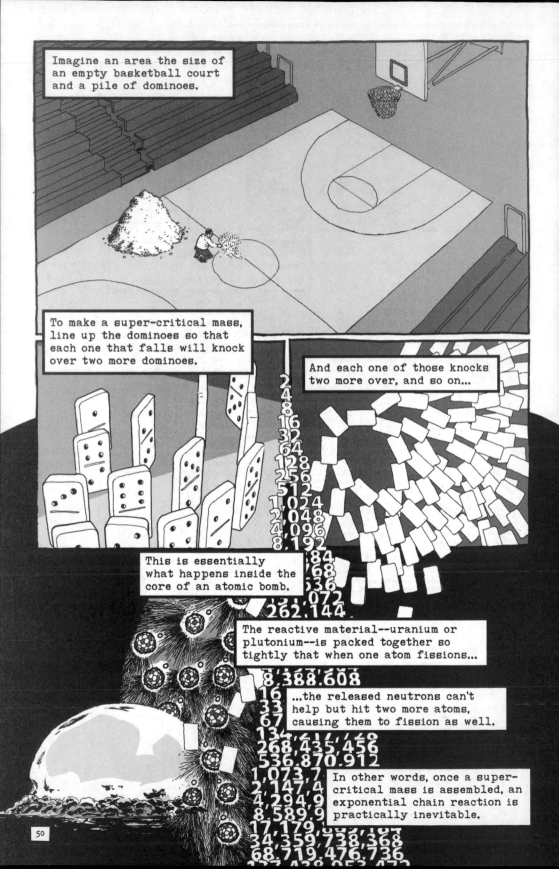

Imagine an area the size of an empty basketball court and a pile of dominoes.

To make a super-critical mass, line up the dominoes so that each one that falls will knock over two more dominoes.

And each one of those knocks two more over, and so on...

This is essentially what happens inside the core of an atomic bomb.

The reactive material--uranium or plutonium--is packed together so tightly that when one atom fissions...

...the released neutrons can't help but hit two more atoms, causing them to fission as well.

In other words, once a super-critical mass is assembled, an exponential chain reaction is practically inevitable.

2
4
8
16
32
64
128
256
512
1,024
2,048
4,096
8,192

...84
...68
...36
...072
262,144

8,388,608
16
33
67
134
268,435,456
536,870,912
1,073,7
2,147,4
4,294,9
8,589,9
17,179,
34,359,738,368
68,719,476,736

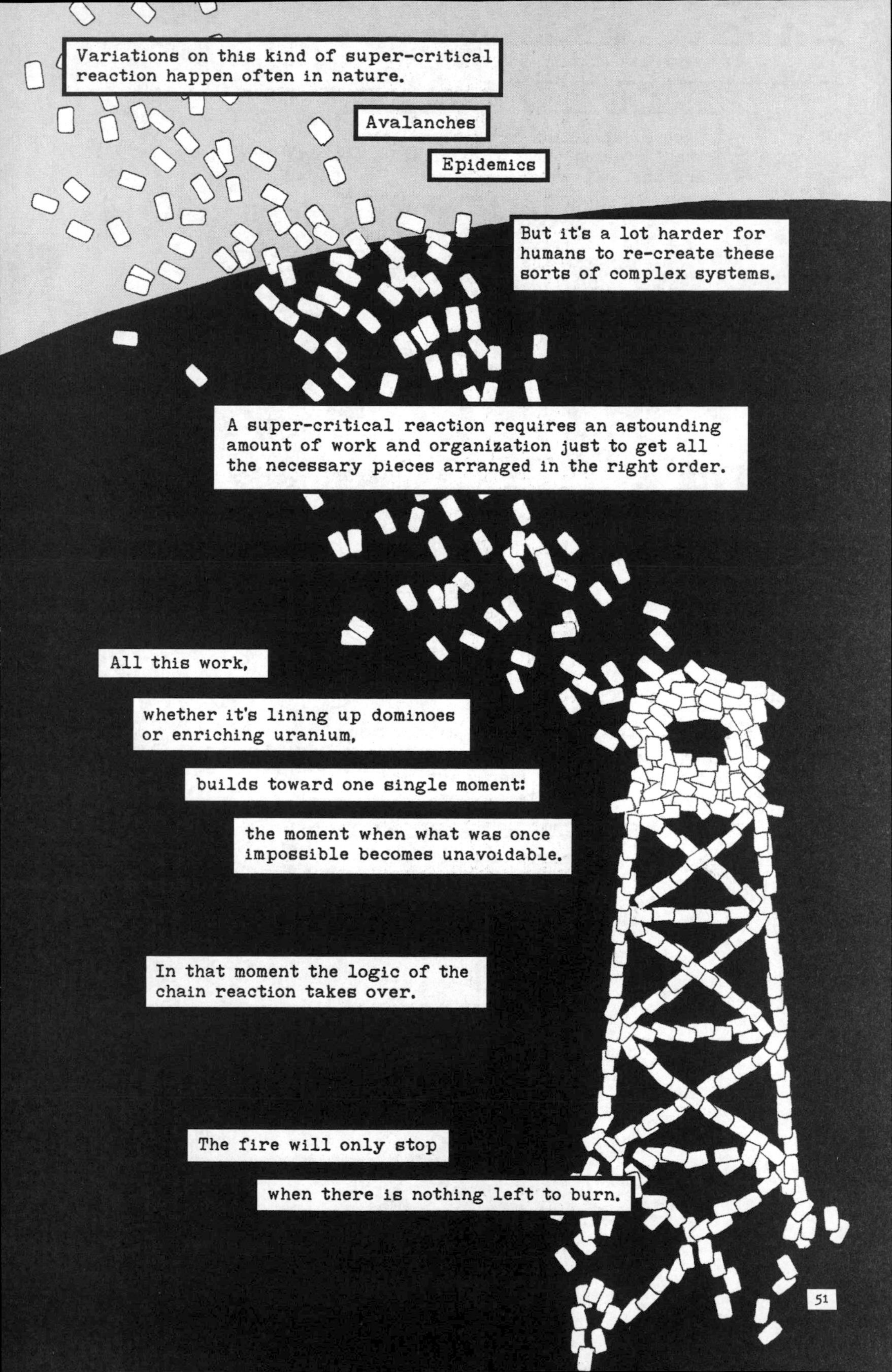

Variations on this kind of super-critical reaction happen often in nature.

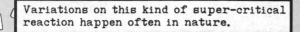

Avalanches

Epidemics

But it's a lot harder for humans to re-create these sorts of complex systems.

A super-critical reaction requires an astounding amount of work and organization just to get all the necessary pieces arranged in the right order.

All this work,

whether it's lining up dominoes or enriching uranium,

builds toward one single moment:

the moment when what was once impossible becomes unavoidable.

In that moment the logic of the chain reaction takes over.

The fire will only stop

when there is nothing left to burn.

The Trinity test was that moment.

Once construction had finished on the factories, the laboratories, and the test sites...

...once the nation's brightest minds had demonstrated the potential power of nuclear fission...

...and, finally, once the military had organized these many parts into a coherent plan to test a bomb...

...a chain reaction was about to be set in motion, making certain outcomes inevitable.

$$N + \operatorname{div}(j) = \frac{v-1}{\tau} N$$
$$j = -D \operatorname{grad} N$$
$$\dot{N} = D \Delta N + \frac{v-1}{\tau} N$$
$$R_c^2 \quad N = N_1(x,y,z) e^{t}$$
$$M_c \Delta N_1 + \frac{-v' + v - 1}{D\tau} N_1 = 0$$
$$\sigma + (v-1) N_1(v) = \frac{\sin(\pi r/R)}{r}$$
$$' = (v-1) - \pi^2 D r/R^2 \qquad \pi^2 D$$
$$e(v-1)t[1-(R_o/R)^2 = \frac{\pi^2 D\tau}{v-1} \quad \frac{\pi^2}{2} |2$$
$$\sigma_t = [\sigma_f + \iint \sigma_s (1-\cos$$
$$_t = 4 \cdot 10^{-24} cm^2$$
$$\frac{1}{\sigma_{f'}} = \frac{1}{V} \frac{\sigma_t}{\sigma_f}$$

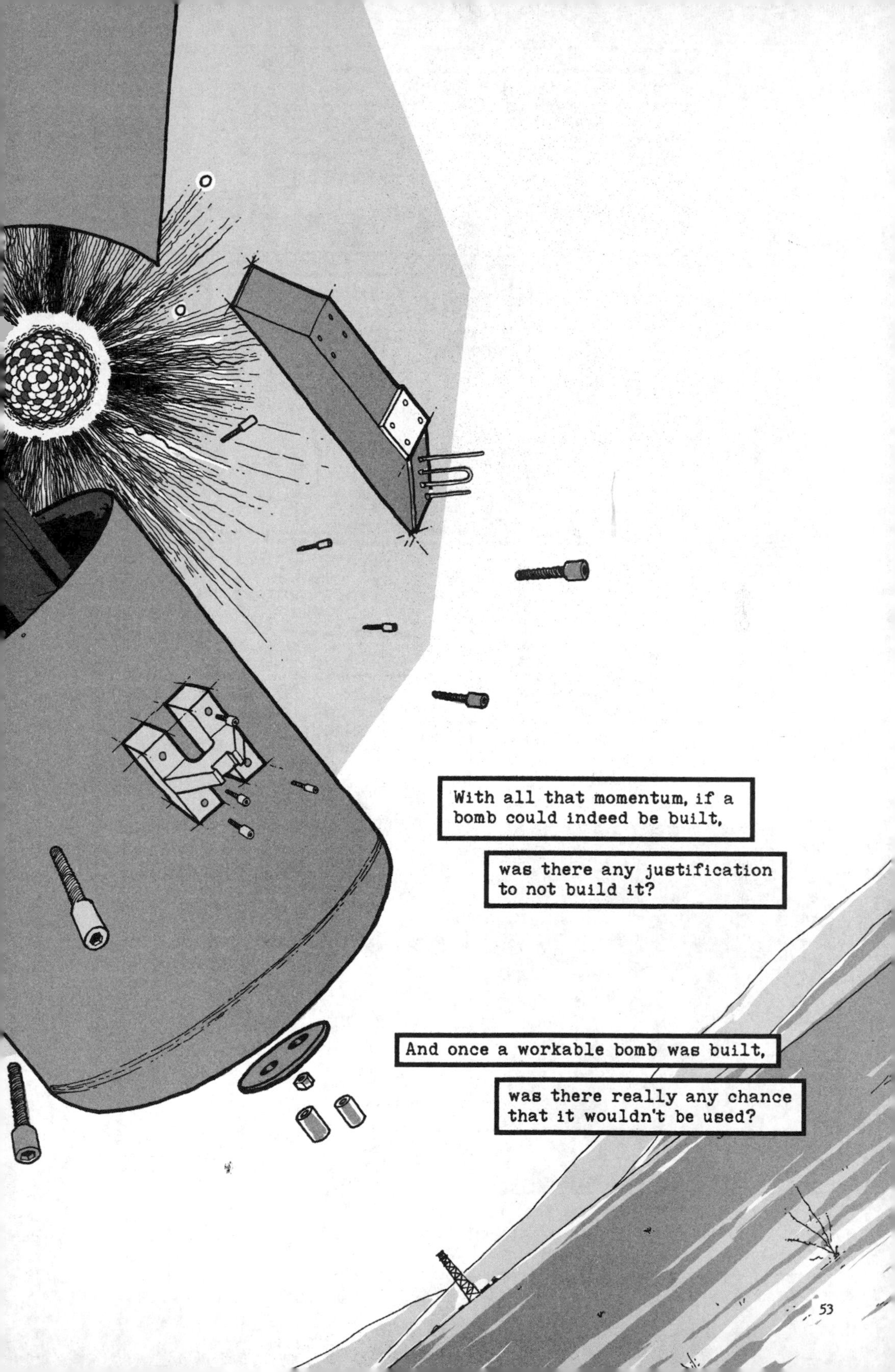
With all that momentum, if a bomb could indeed be built,
was there any justification to not build it?
And once a workable bomb was built,
was there really any chance that it wouldn't be used?

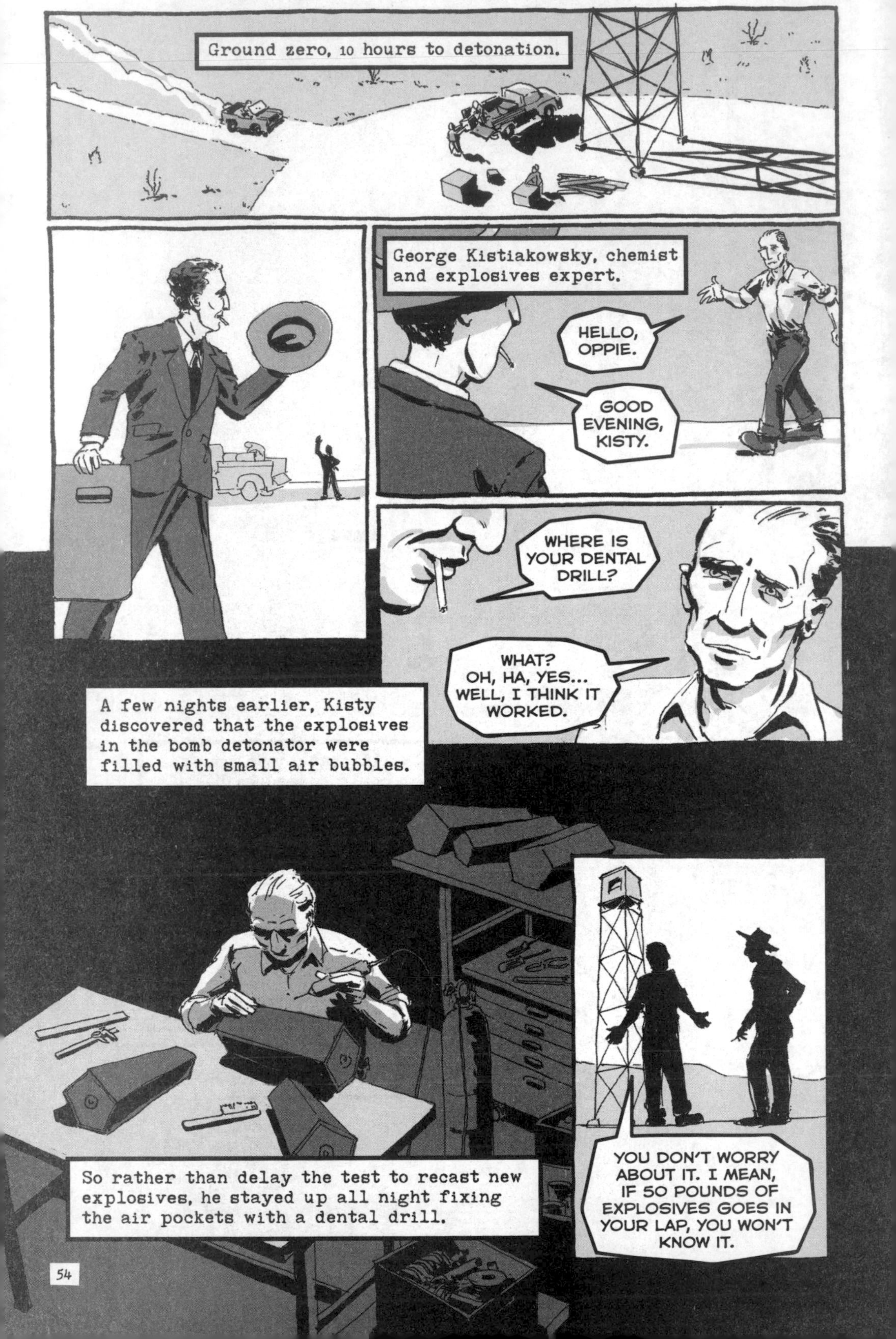
Ground zero, 10 hours to detonation.
George Kistiakowsky, chemist and explosives expert.
HELLO, OPPIE.
GOOD EVENING, KISTY.
WHERE IS YOUR DENTAL DRILL?
WHAT? OH, HA, YES... WELL, I THINK IT WORKED.
A few nights earlier, Kisty discovered that the explosives in the bomb detonator were filled with small air bubbles.
So rather than delay the test to recast new explosives, he stayed up all night fixing the air pockets with a dental drill.
YOU DON'T WORRY ABOUT IT. I MEAN, IF 50 POUNDS OF EXPLOSIVES GOES IN YOUR LAP, YOU WON'T KNOW IT.

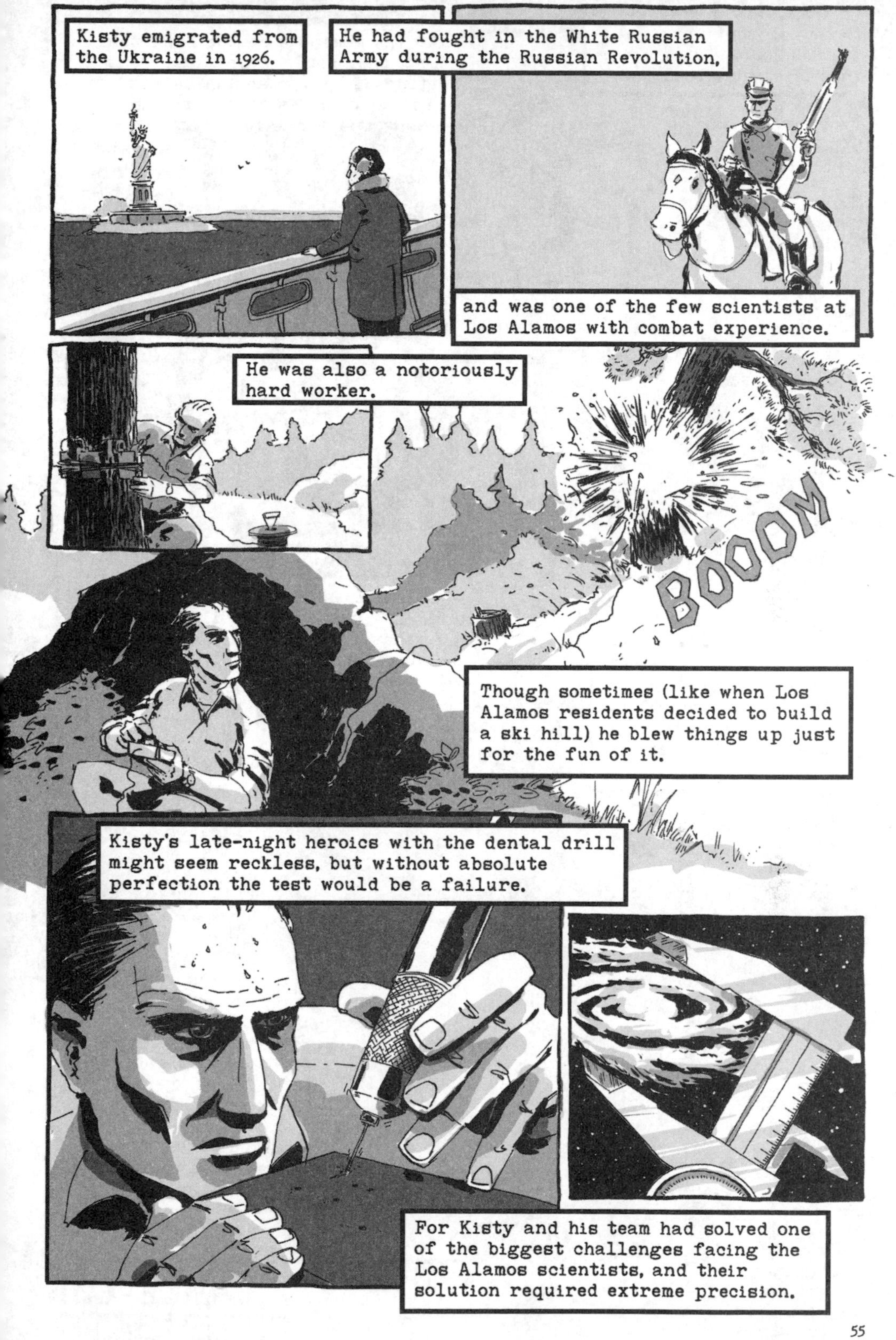
Kisty emigrated from the Ukraine in 1926.
He had fought in the White Russian Army during the Russian Revolution,
and was one of the few scientists at Los Alamos with combat experience.
He was also a notoriously hard worker.
BOOOM
Though sometimes (like when Los Alamos residents decided to build a ski hill) he blew things up just for the fun of it.
Kisty's late-night heroics with the dental drill might seem reckless, but without absolute perfection the test would be a failure.
For Kisty and his team had solved one of the biggest challenges facing the Los Alamos scientists, and their solution required extreme precision.

Remember that uranium 235 and plutonium 239 are incredibly unstable.

Like the dominoes on the basketball court, they could topple accidentally, setting off the reaction before the complete chain is even assembled.

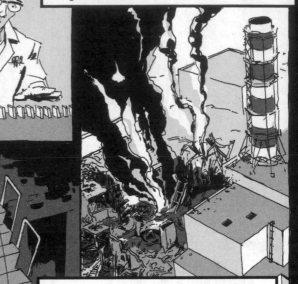

With dominoes this would just be annoying, but with fissile material a premature chain reaction would be devastating.

Because neutrons are whizzing around us all the time, there's always a chance that a stray one might start a chain reaction.

NO!

So bomb designers had to figure out a way to keep the fissile material packed closely together inside a bomb casing,

but without the material being close enough to form a super-critical mass.

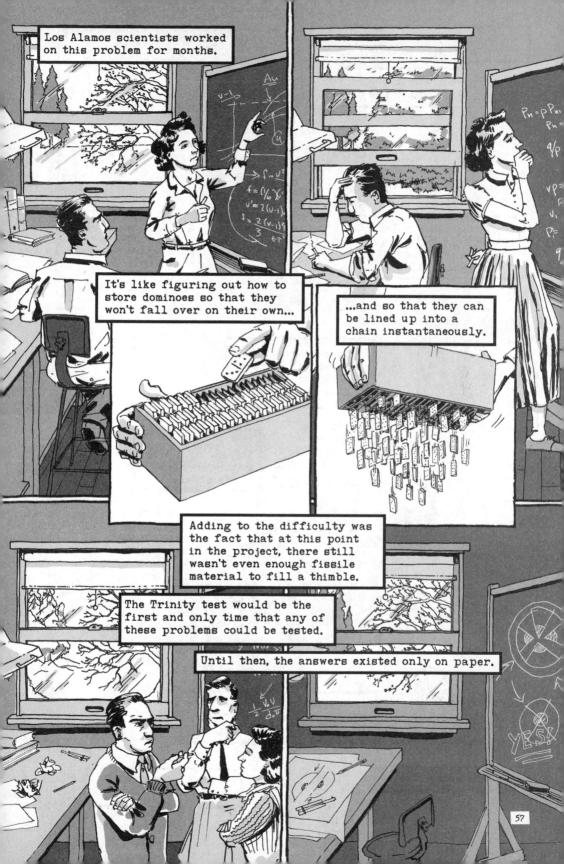

Los Alamos scientists worked on this problem for months.

It's like figuring out how to store dominoes so that they won't fall over on their own...

...and so that they can be lined up into a chain instantaneously.

Adding to the difficulty was the fact that at this point in the project, there still wasn't even enough fissile material to fill a thimble.

The Trinity test would be the first and only time that any of these problems could be tested.

Until then, the answers existed only on paper.

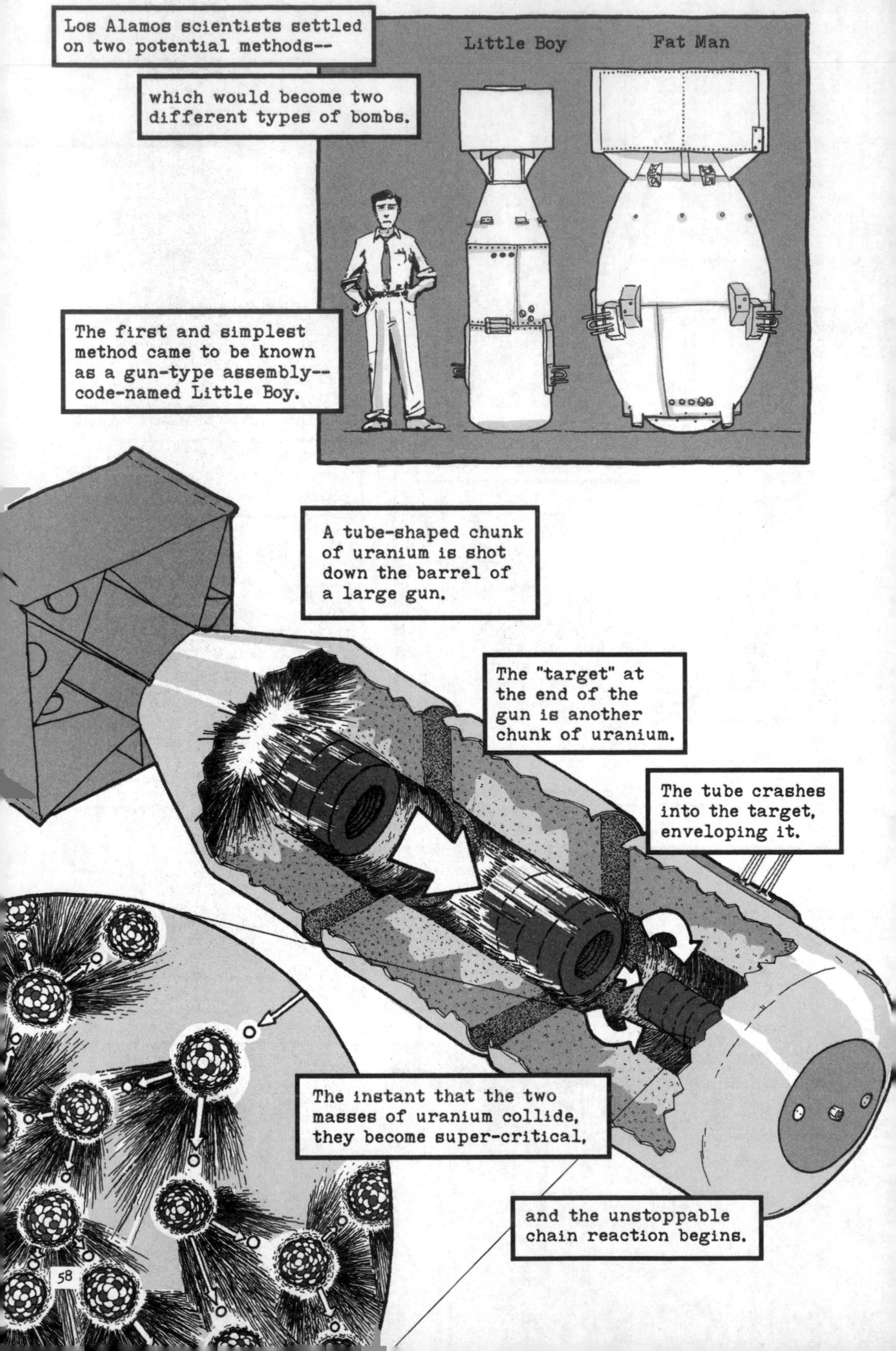
Los Alamos scientists settled on two potential methods--
which would become two different types of bombs.
Little Boy
Fat Man
The first and simplest method came to be known as a gun-type assembly--code-named Little Boy.
A tube-shaped chunk of uranium is shot down the barrel of a large gun.
The "target" at the end of the gun is another chunk of uranium.
The tube crashes into the target, enveloping it.
The instant that the two masses of uranium collide, they become super-critical,
and the unstoppable chain reaction begins.

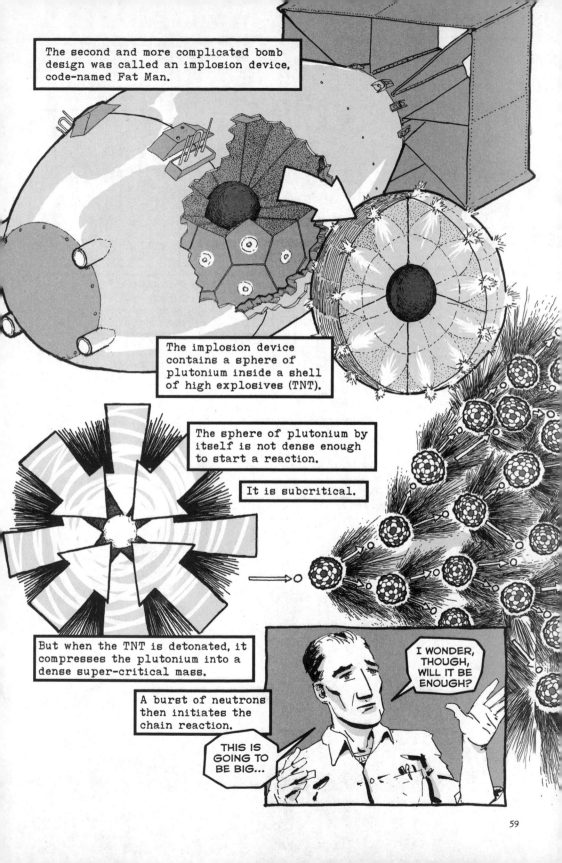

The second and more complicated bomb design was called an implosion device, code-named Fat Man.

The implosion device contains a sphere of plutonium inside a shell of high explosives (TNT).

The sphere of plutonium by itself is not dense enough to start a reaction.

It is subcritical.

But when the TNT is detonated, it compresses the plutonium into a dense super-critical mass.

A burst of neutrons then initiates the chain reaction.

THIS IS GOING TO BE BIG...

I WONDER, THOUGH, WILL IT BE ENOUGH?

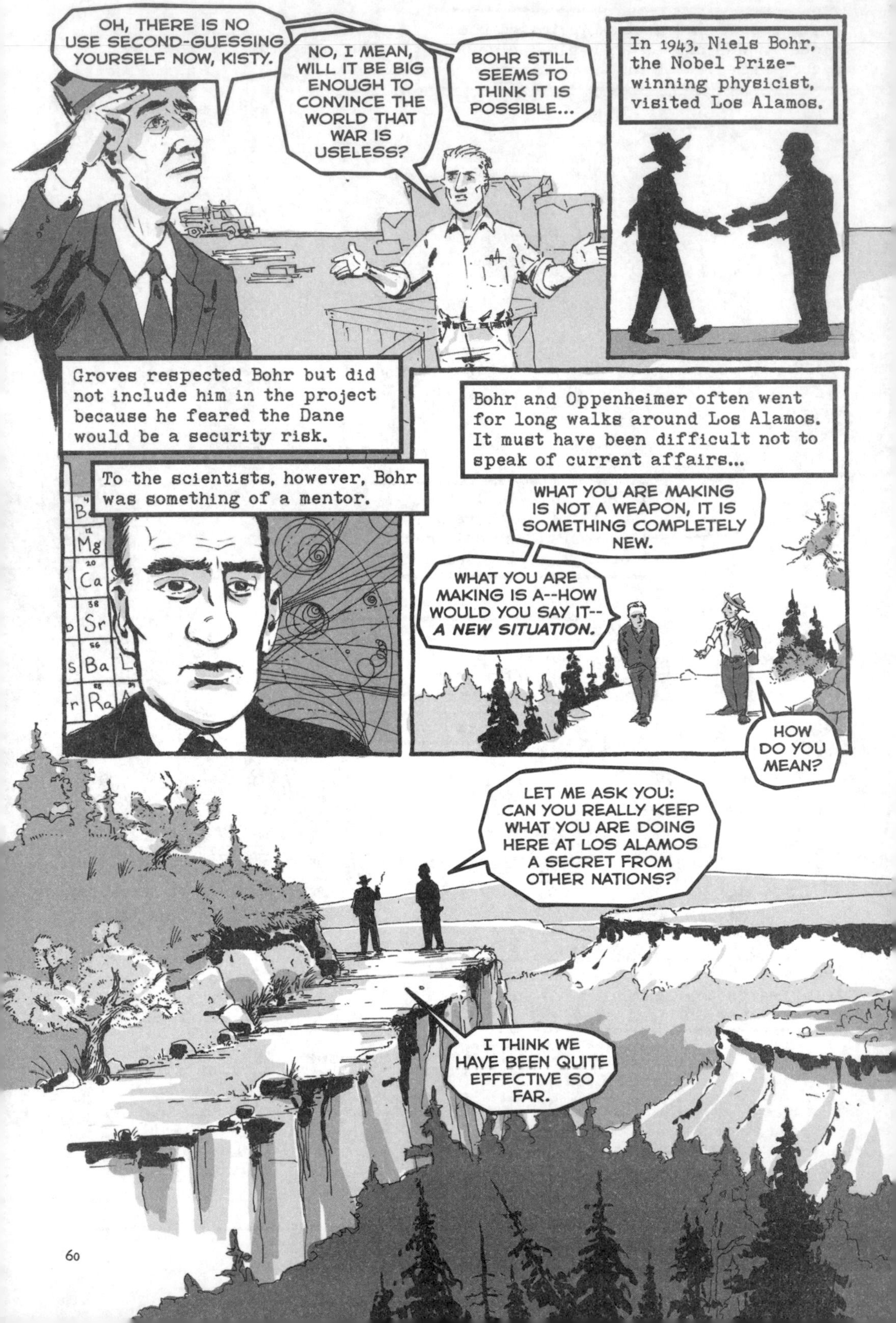
OH, THERE IS NO USE SECOND-GUESSING YOURSELF NOW, KISTY.
NO, I MEAN, WILL IT BE BIG ENOUGH TO CONVINCE THE WORLD THAT WAR IS USELESS?
BOHR STILL SEEMS TO THINK IT IS POSSIBLE...
In 1943, Niels Bohr, the Nobel Prize-winning physicist, visited Los Alamos.
Groves respected Bohr but did not include him in the project because he feared the Dane would be a security risk.
To the scientists, however, Bohr was something of a mentor.
Bohr and Oppenheimer often went for long walks around Los Alamos. It must have been difficult not to speak of current affairs...
WHAT YOU ARE MAKING IS NOT A WEAPON, IT IS SOMETHING COMPLETELY NEW.
WHAT YOU ARE MAKING IS A--HOW WOULD YOU SAY IT-- *A NEW SITUATION.*
HOW DO YOU MEAN?
LET ME ASK YOU: CAN YOU REALLY KEEP WHAT YOU ARE DOING HERE AT LOS ALAMOS A SECRET FROM OTHER NATIONS?
I THINK WE HAVE BEEN QUITE EFFECTIVE SO FAR.

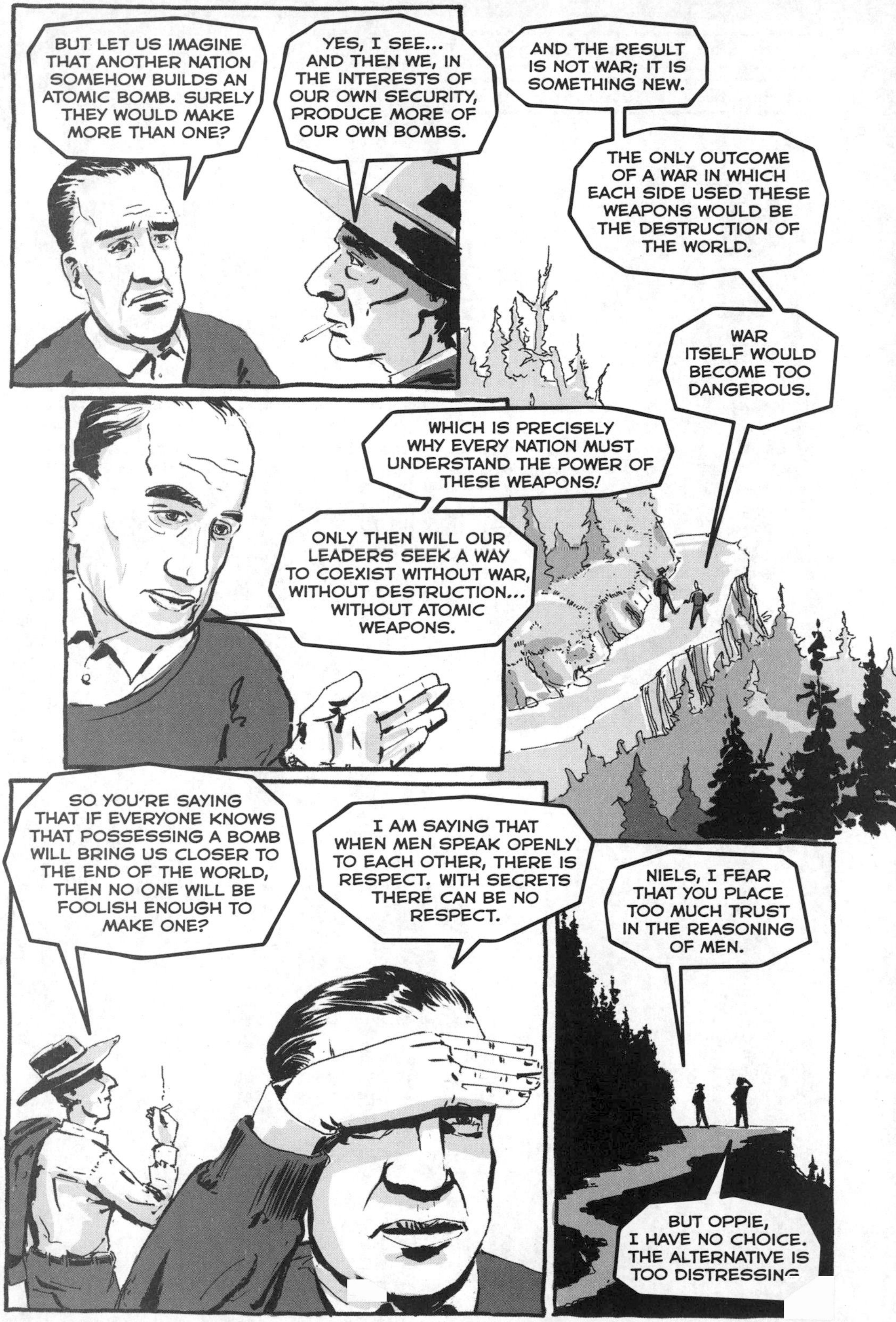

BUT LET US IMAGINE THAT ANOTHER NATION SOMEHOW BUILDS AN ATOMIC BOMB. SURELY THEY WOULD MAKE MORE THAN ONE?
YES, I SEE... AND THEN WE, IN THE INTERESTS OF OUR OWN SECURITY, PRODUCE MORE OF OUR OWN BOMBS.
AND THE RESULT IS NOT WAR; IT IS SOMETHING NEW.
THE ONLY OUTCOME OF A WAR IN WHICH EACH SIDE USED THESE WEAPONS WOULD BE THE DESTRUCTION OF THE WORLD.
WAR ITSELF WOULD BECOME TOO DANGEROUS.
WHICH IS PRECISELY WHY EVERY NATION MUST UNDERSTAND THE POWER OF THESE WEAPONS!
ONLY THEN WILL OUR LEADERS SEEK A WAY TO COEXIST WITHOUT WAR, WITHOUT DESTRUCTION... WITHOUT ATOMIC WEAPONS.
SO YOU'RE SAYING THAT IF EVERYONE KNOWS THAT POSSESSING A BOMB WILL BRING US CLOSER TO THE END OF THE WORLD, THEN NO ONE WILL BE FOOLISH ENOUGH TO MAKE ONE?
I AM SAYING THAT WHEN MEN SPEAK OPENLY TO EACH OTHER, THERE IS RESPECT. WITH SECRETS THERE CAN BE NO RESPECT.
NIELS, I FEAR THAT YOU PLACE TOO MUCH TRUST IN THE REASONING OF MEN.
BUT OPPIE, I HAVE NO CHOICE. THE ALTERNATIVE IS TOO DISTRESSIN

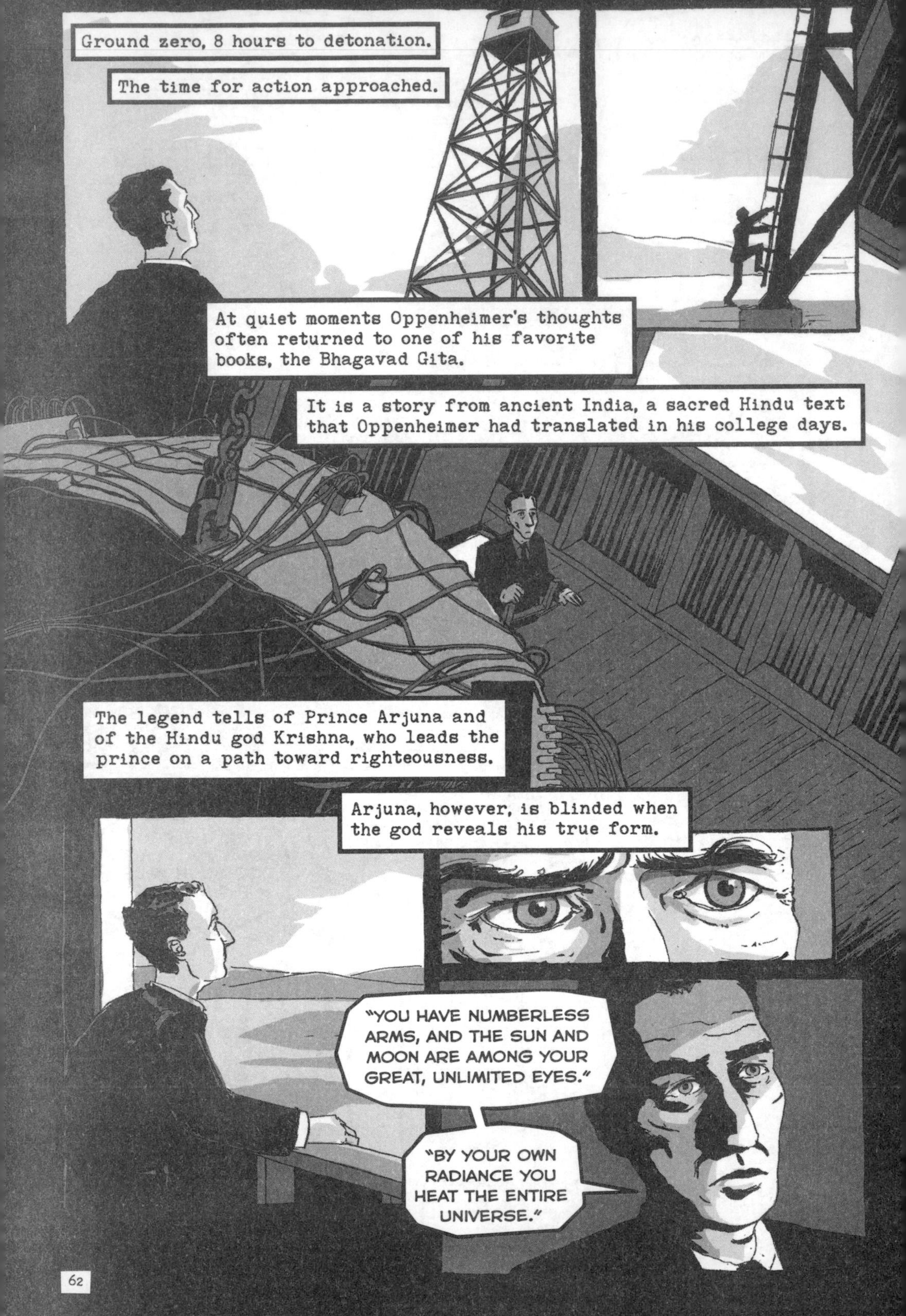

Ground zero, 8 hours to detonation.
The time for action approached.
At quiet moments Oppenheimer's thoughts often returned to one of his favorite books, the Bhagavad Gita.
It is a story from ancient India, a sacred Hindu text that Oppenheimer had translated in his college days.
The legend tells of Prince Arjuna and of the Hindu god Krishna, who leads the prince on a path toward righteousness.
Arjuna, however, is blinded when the god reveals his true form.
"YOU HAVE NUMBERLESS ARMS, AND THE SUN AND MOON ARE AMONG YOUR GREAT, UNLIMITED EYES."
"BY YOUR OWN RADIANCE YOU HEAT THE ENTIRE UNIVERSE."

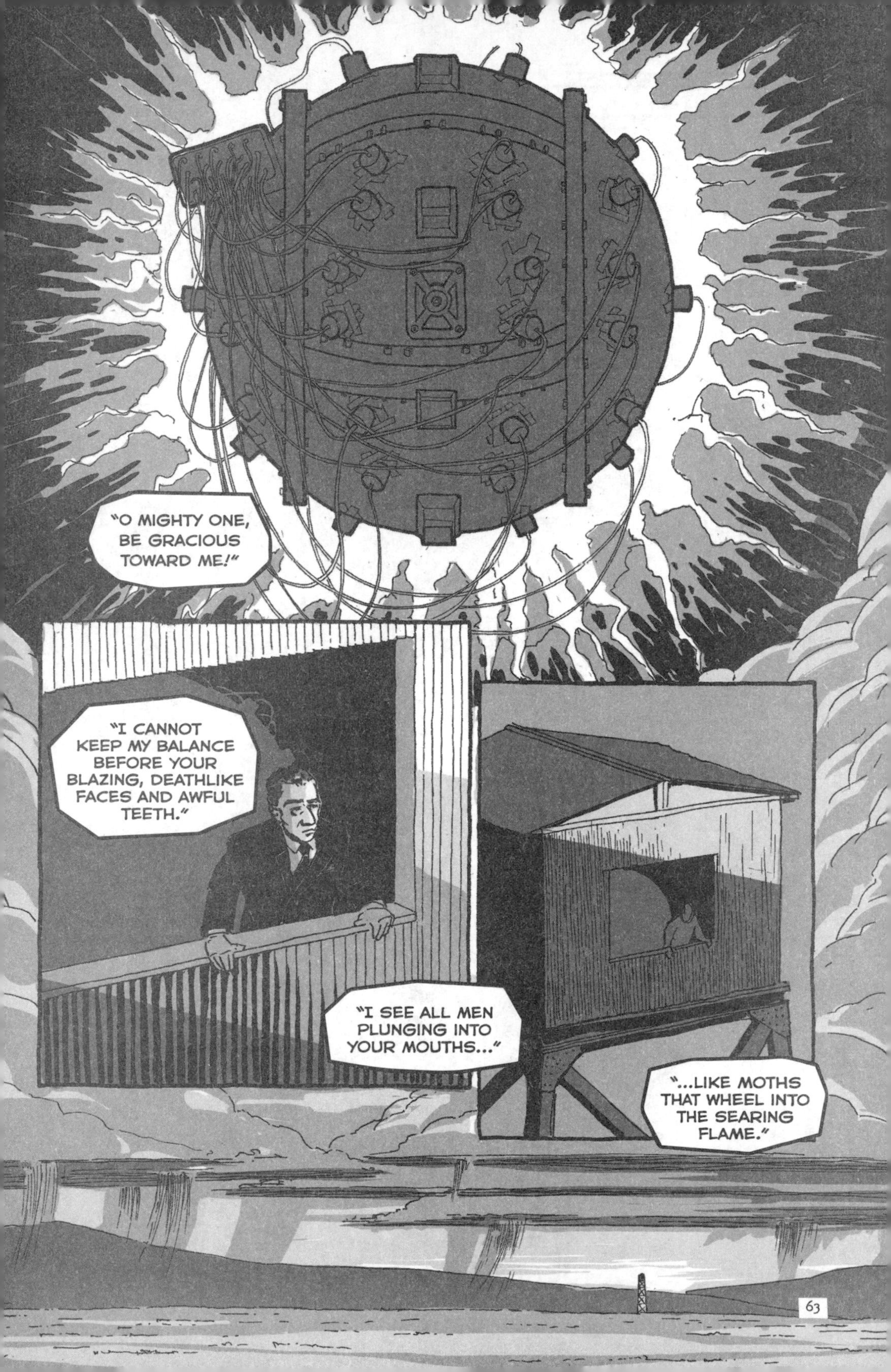
"O MIGHTY ONE, BE GRACIOUS TOWARD ME!"
"I CANNOT KEEP MY BALANCE BEFORE YOUR BLAZING, DEATHLIKE FACES AND AWFUL TEETH."
"I SEE ALL MEN PLUNGING INTO YOUR MOUTHS..."
"...LIKE MOTHS THAT WHEEL INTO THE SEARING FLAME."

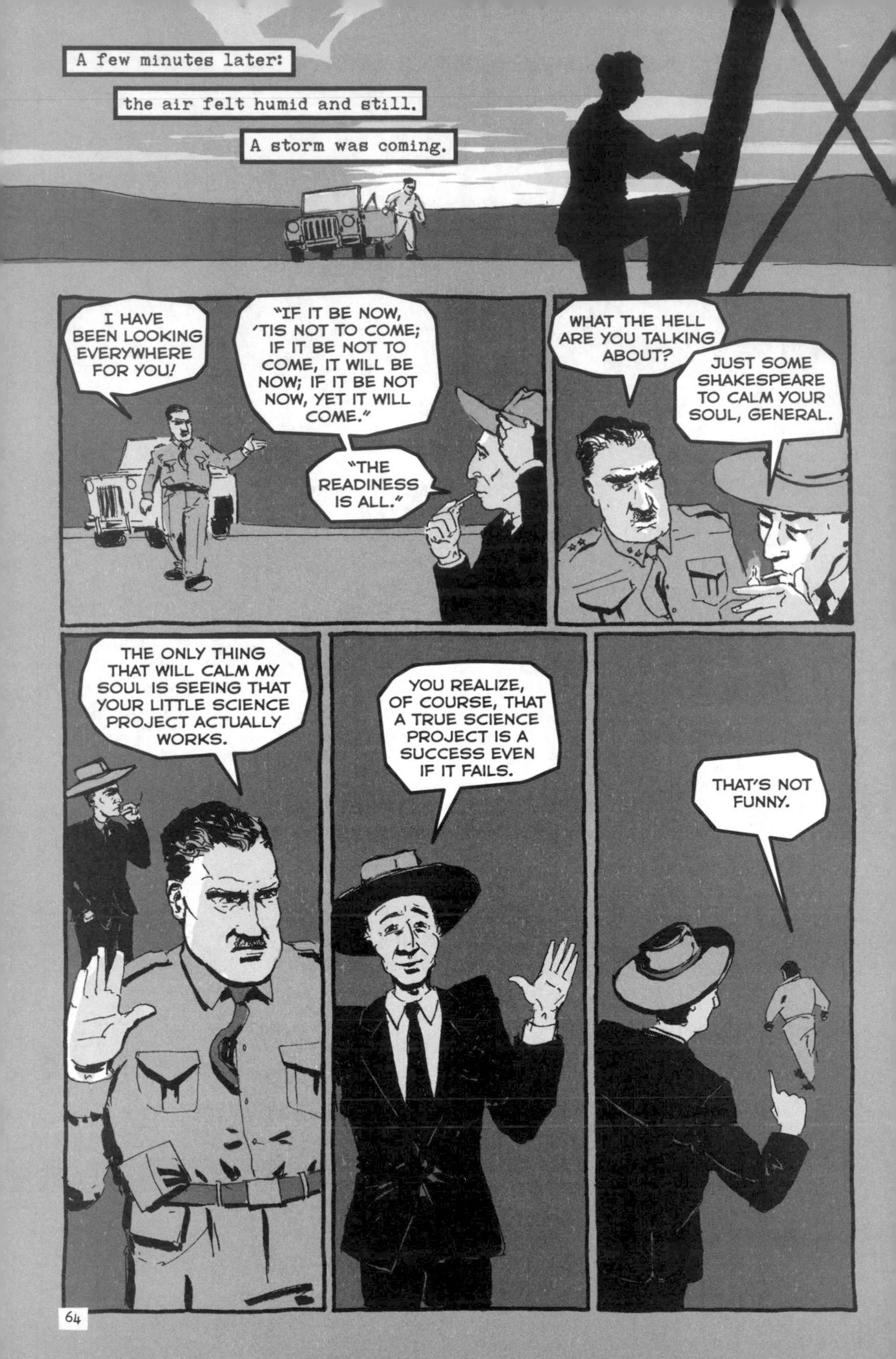
A few minutes later:
the air felt humid and still.
A storm was coming.
I HAVE BEEN LOOKING EVERYWHERE FOR YOU!
"IF IT BE NOW, 'TIS NOT TO COME; IF IT BE NOT TO COME, IT WILL BE NOW; IF IT BE NOT NOW, YET IT WILL COME."
"THE READINESS IS ALL."
WHAT THE HELL ARE YOU TALKING ABOUT?
JUST SOME SHAKESPEARE TO CALM YOUR SOUL, GENERAL.
THE ONLY THING THAT WILL CALM MY SOUL IS SEEING THAT YOUR LITTLE SCIENCE PROJECT ACTUALLY WORKS.
YOU REALIZE, OF COURSE, THAT A TRUE SCIENCE PROJECT IS A SUCCESS EVEN IF IT FAILS.
THAT'S NOT FUNNY.

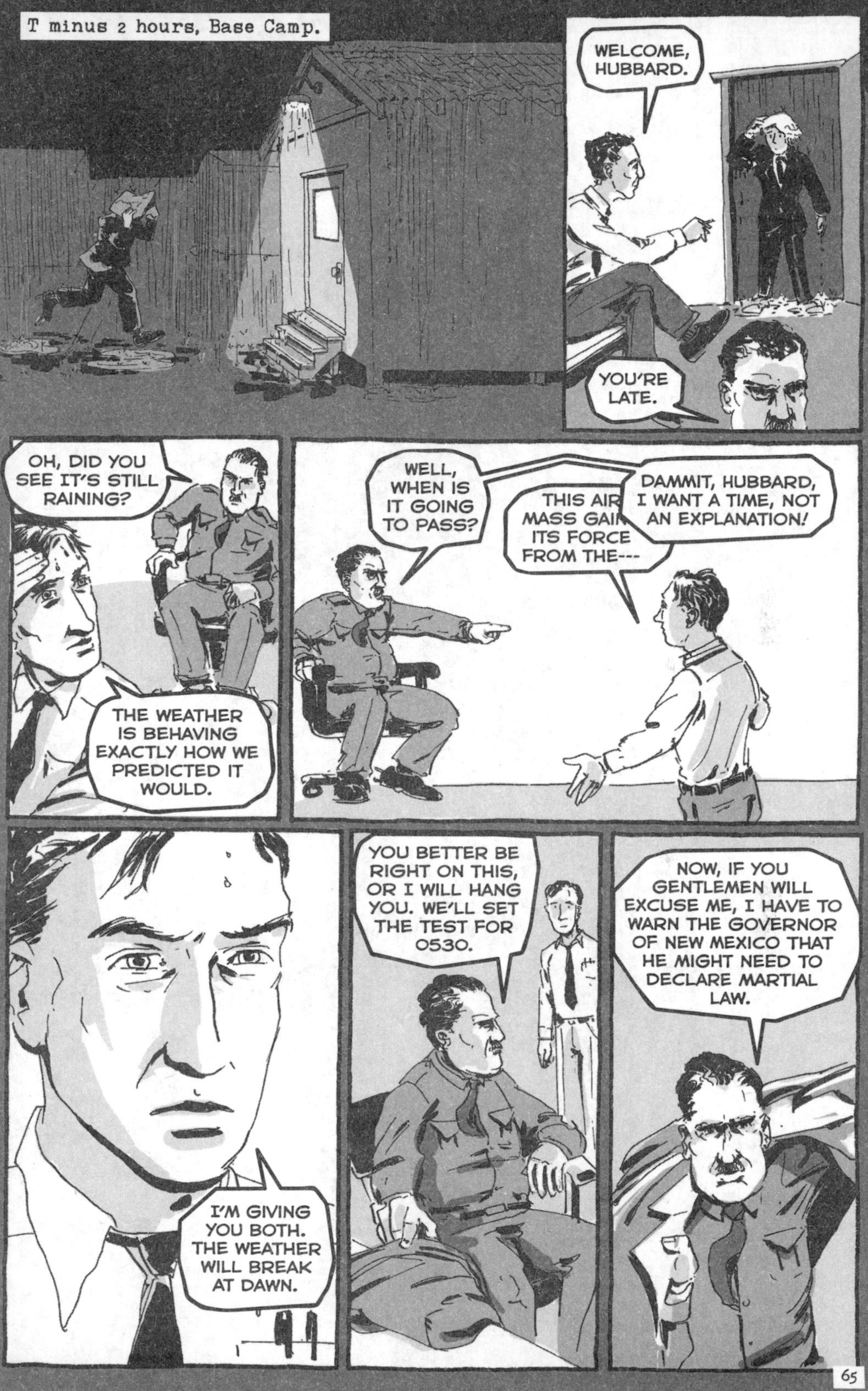
T minus 2 hours, Base Camp.
WELCOME, HUBBARD.
YOU'RE LATE.
OH, DID YOU SEE IT'S STILL RAINING?
THE WEATHER IS BEHAVING EXACTLY HOW WE PREDICTED IT WOULD.
WELL, WHEN IS IT GOING TO PASS?
THIS AIR MASS GAIN ITS FORCE FROM THE---
DAMMIT, HUBBARD, I WANT A TIME, NOT AN EXPLANATION!
I'M GIVING YOU BOTH. THE WEATHER WILL BREAK AT DAWN.
YOU BETTER BE RIGHT ON THIS, OR I WILL HANG YOU. WE'LL SET THE TEST FOR 0530.
NOW, IF YOU GENTLEMEN WILL EXCUSE ME, I HAVE TO WARN THE GOVERNOR OF NEW MEXICO THAT HE MIGHT NEED TO DECLARE MARTIAL LAW.

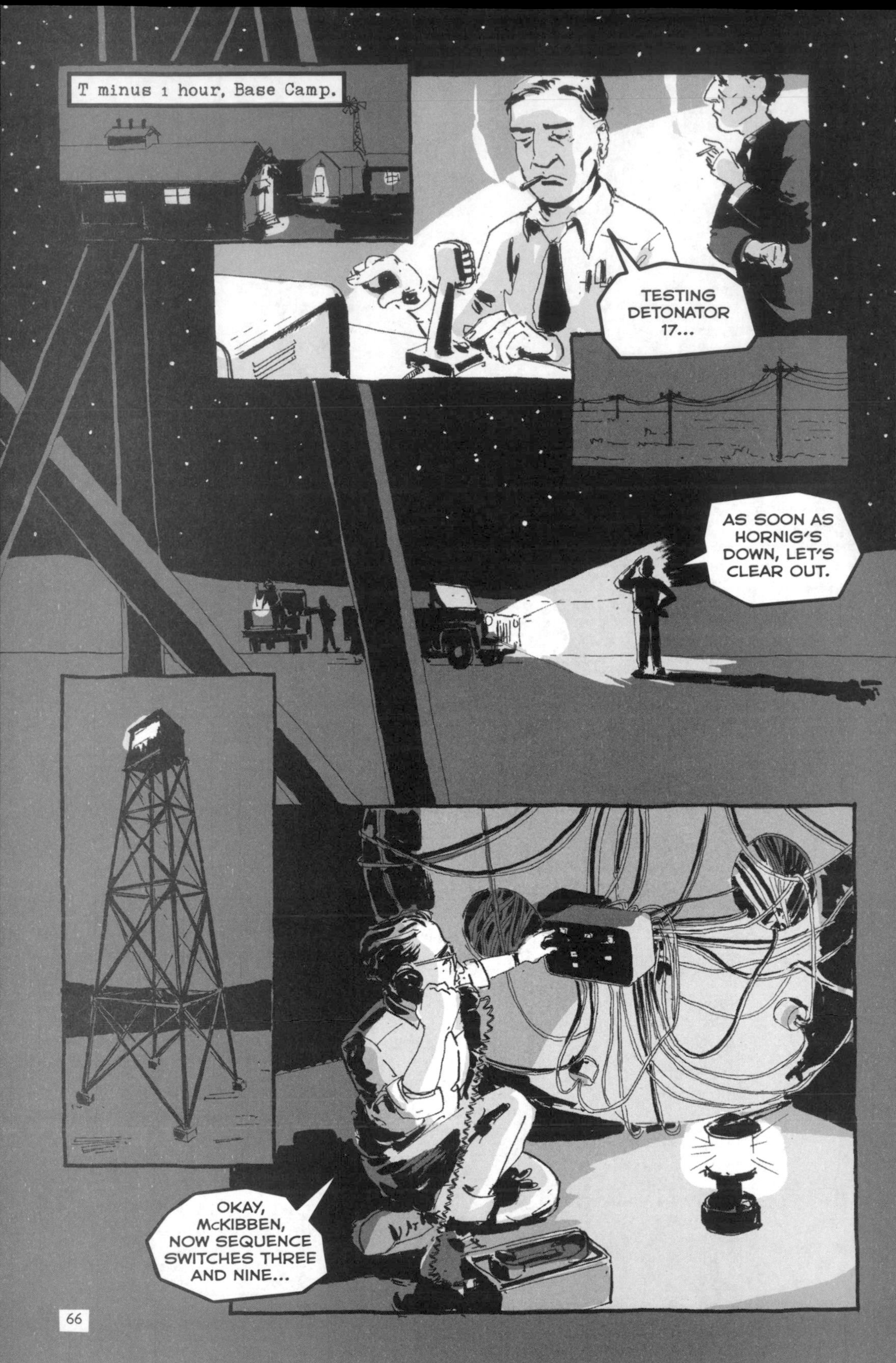

T minus 1 hour, Base Camp.
TESTING DETONATOR 17...
AS SOON AS HORNIG'S DOWN, LET'S CLEAR OUT.
OKAY, McKIBBEN, NOW SEQUENCE SWITCHES THREE AND NINE...

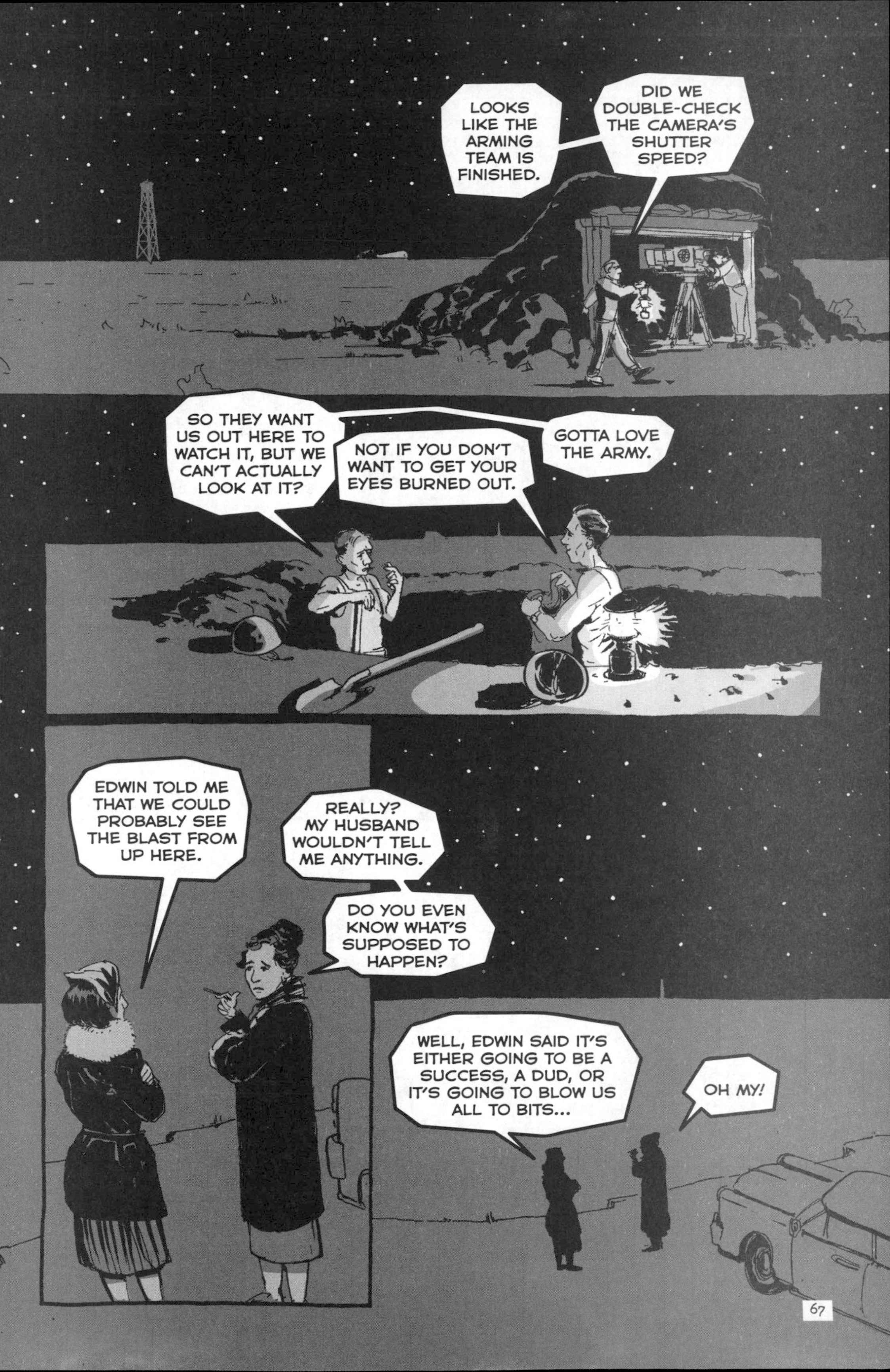
LOOKS LIKE THE ARMING TEAM IS FINISHED.
DID WE DOUBLE-CHECK THE CAMERA'S SHUTTER SPEED?
SO THEY WANT US OUT HERE TO WATCH IT, BUT WE CAN'T ACTUALLY LOOK AT IT?
NOT IF YOU DON'T WANT TO GET YOUR EYES BURNED OUT.
GOTTA LOVE THE ARMY.
EDWIN TOLD ME THAT WE COULD PROBABLY SEE THE BLAST FROM UP HERE.
REALLY? MY HUSBAND WOULDN'T TELL ME ANYTHING.
DO YOU EVEN KNOW WHAT'S SUPPOSED TO HAPPEN?
WELL, EDWIN SAID IT'S EITHER GOING TO BE A SUCCESS, A DUD, OR IT'S GOING TO BLOW US ALL TO BITS...
OH MY!

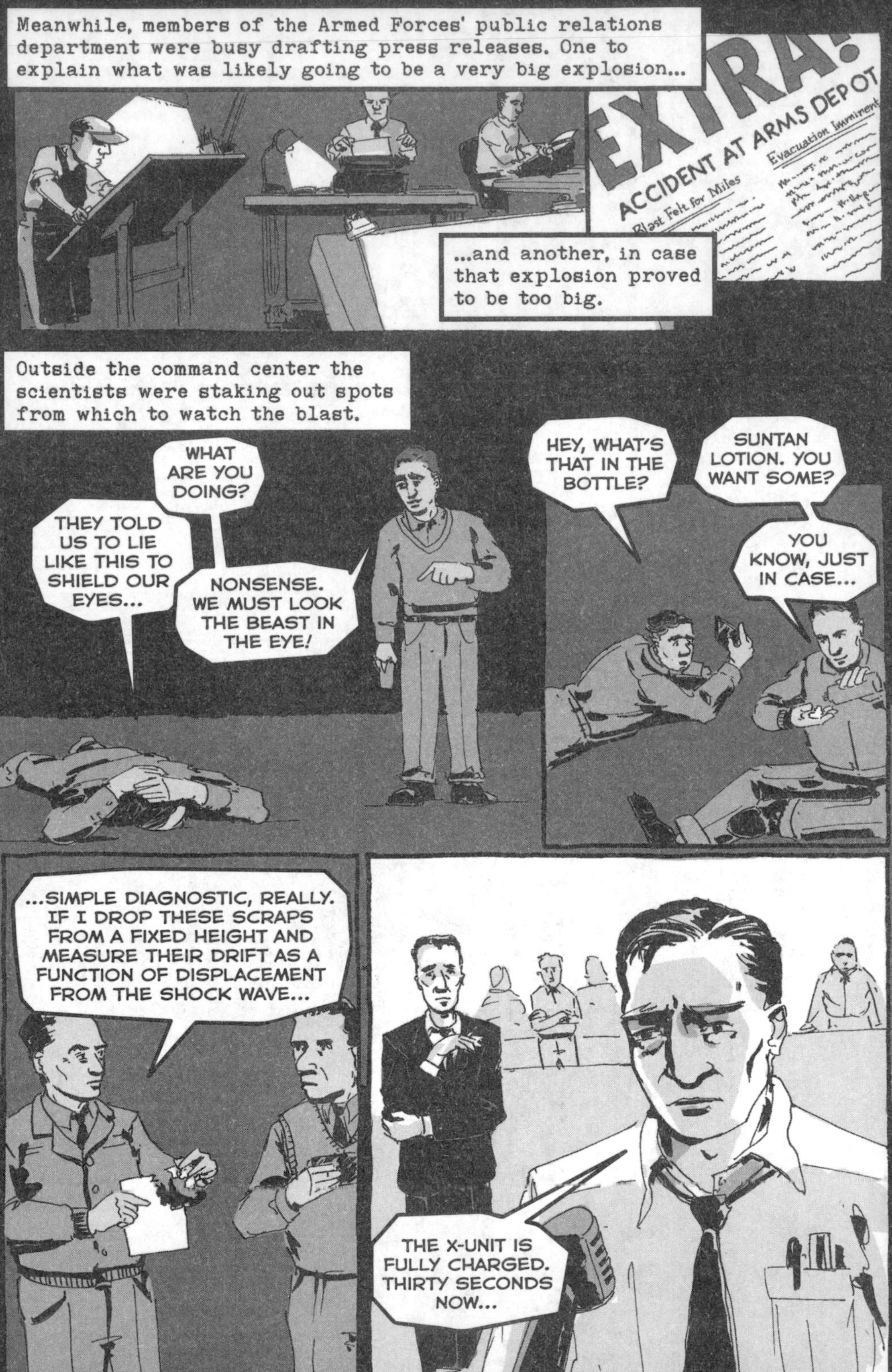
Meanwhile, members of the Armed Forces' public relations department were busy drafting press releases. One to explain what was likely going to be a very big explosion...
EXTRA!
ACCIDENT AT ARMS DEPOT
Blast Felt for Miles
Evacuation Imminent
...and another, in case that explosion proved to be too big.
Outside the command center the scientists were staking out spots from which to watch the blast.
THEY TOLD US TO LIE LIKE THIS TO SHIELD OUR EYES...
WHAT ARE YOU DOING?
NONSENSE. WE MUST LOOK THE BEAST IN THE EYE!
HEY, WHAT'S THAT IN THE BOTTLE?
SUNTAN LOTION. YOU WANT SOME?
YOU KNOW, JUST IN CASE...
...SIMPLE DIAGNOSTIC, REALLY. IF I DROP THESE SCRAPS FROM A FIXED HEIGHT AND MEASURE THEIR DRIFT AS A FUNCTION OF DISPLACEMENT FROM THE SHOCK WAVE...
THE X-UNIT IS FULLY CHARGED. THIRTY SECONDS NOW...

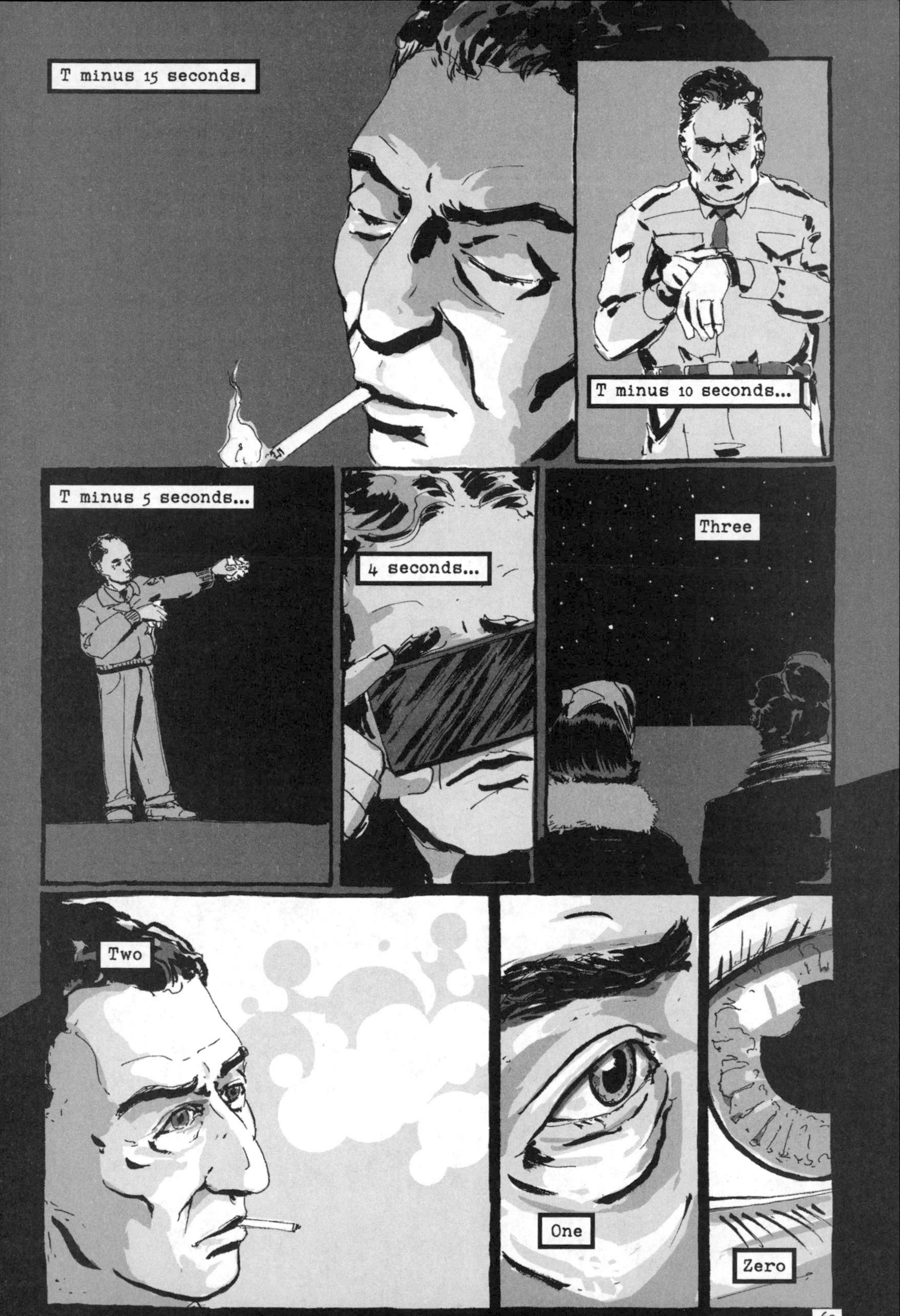
T minus 15 seconds.
T minus 10 seconds...
T minus 5 seconds...
4 seconds...
Three
Two
One
Zero

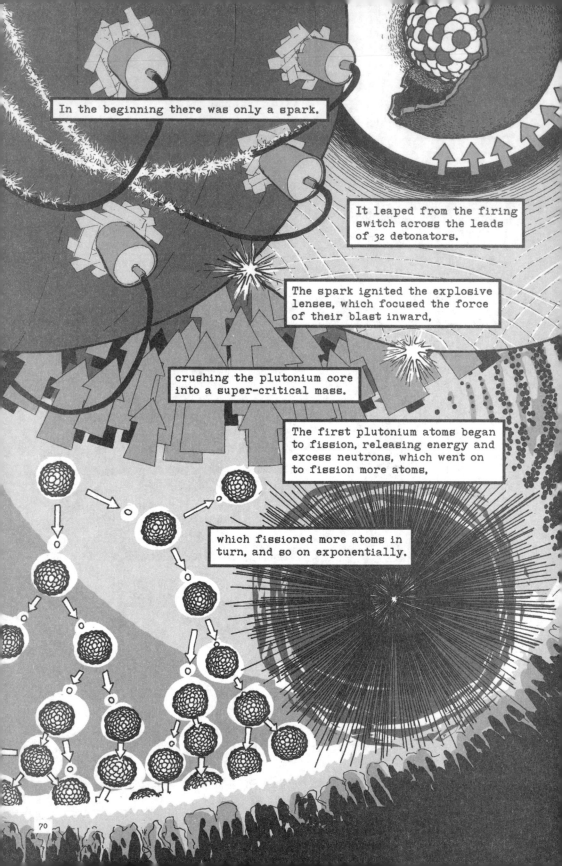

In the beginning there was only a spark.

It leaped from the firing switch across the leads of 32 detonators.

The spark ignited the explosive lenses, which focused the force of their blast inward,

crushing the plutonium core into a super-critical mass.

The first plutonium atoms began to fission, releasing energy and excess neutrons, which went on to fission more atoms,

which fissioned more atoms in turn, and so on exponentially.

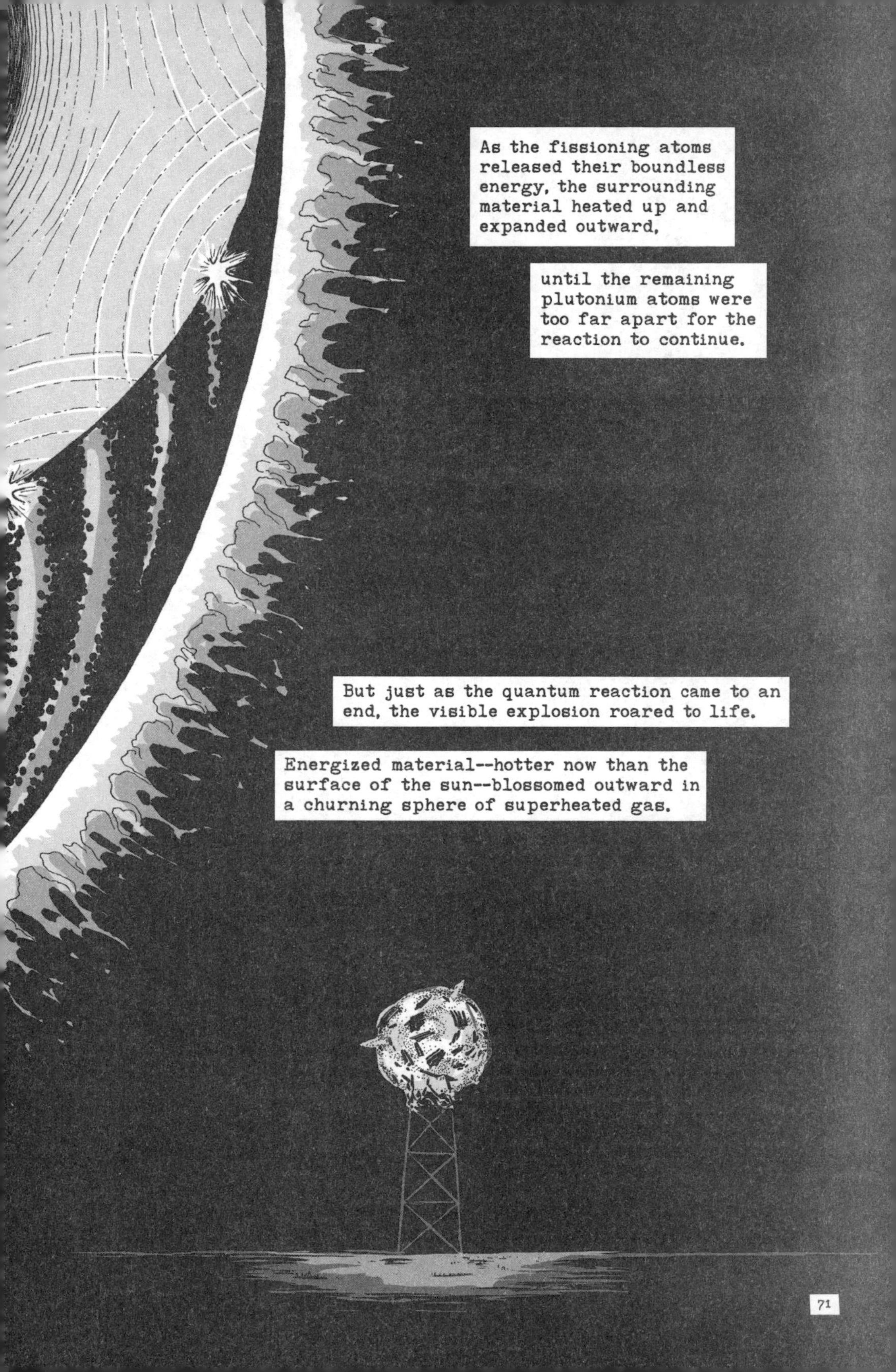
As the fissioning atoms released their boundless energy, the surrounding material heated up and expanded outward,
until the remaining plutonium atoms were too far apart for the reaction to continue.
But just as the quantum reaction came to an end, the visible explosion roared to life.
Energized material--hotter now than the surface of the sun--blossomed outward in a churning sphere of superheated gas.

The rest of the world fell away into darkness.

There was only the light.

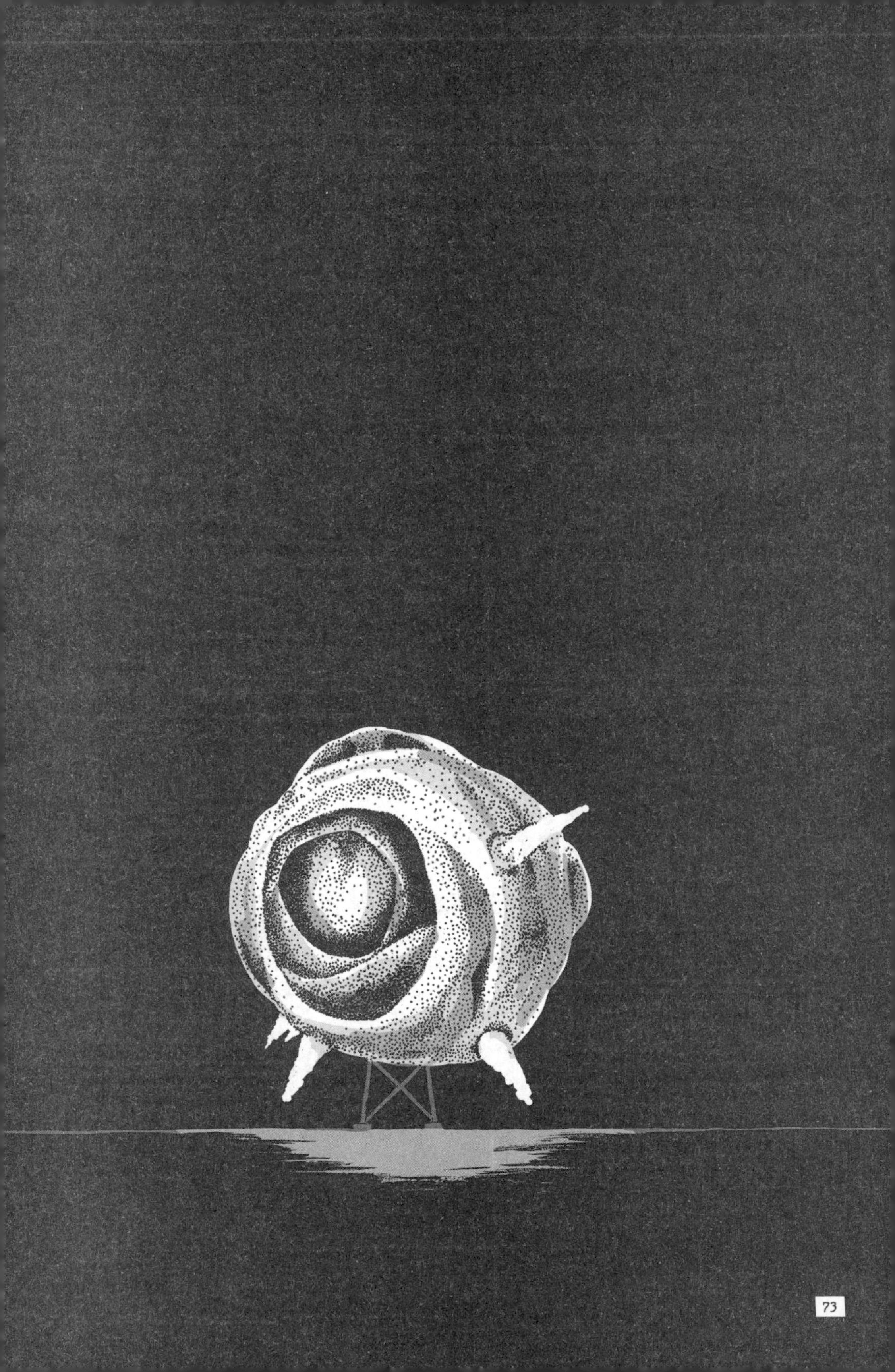

In these fractions of a second,
the atomic light was still pure.

It was not yet a bomb.

It was not yet a symbol of apocalypse.

It was not yet a
part of our world.

In these fractions of a second,

the atomic light was still as timeless
and indifferent as the universe itself.

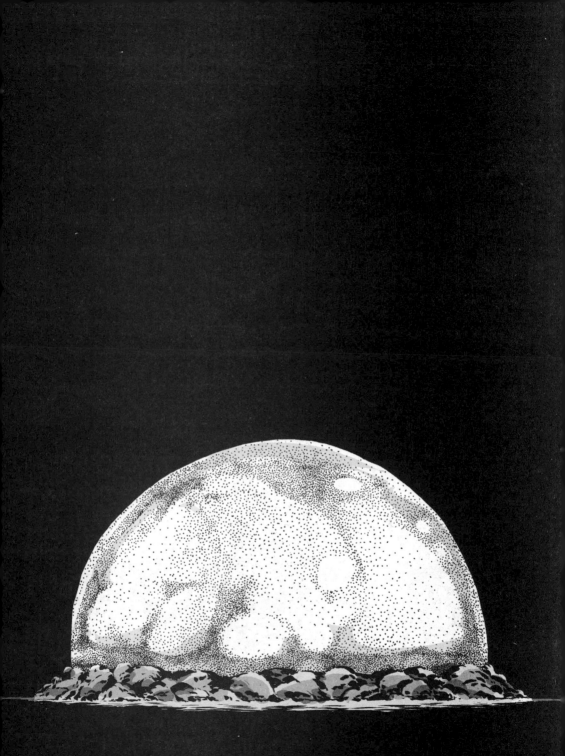

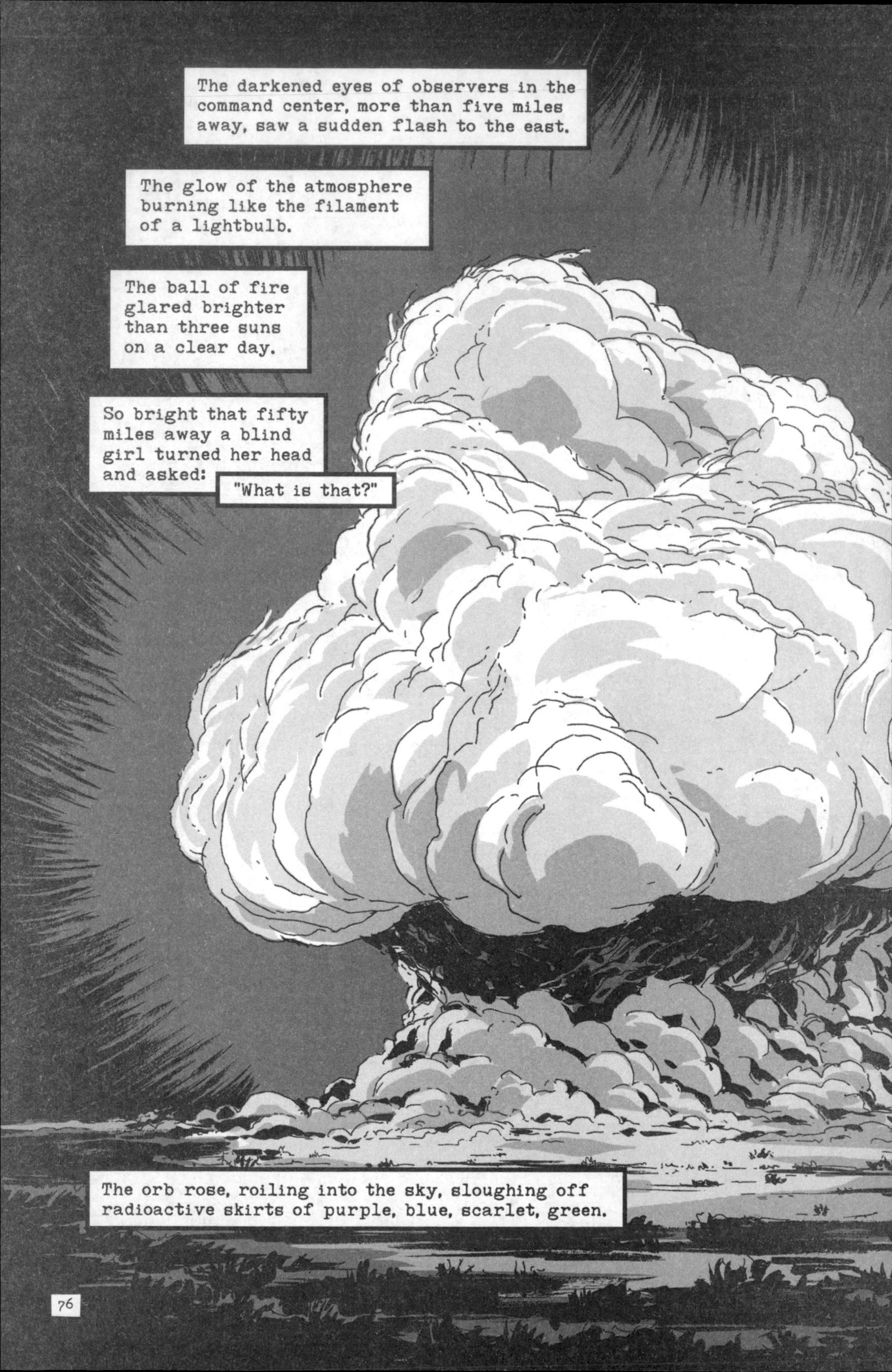
The darkened eyes of observers in the command center, more than five miles away, saw a sudden flash to the east.
The glow of the atmosphere burning like the filament of a lightbulb.
The ball of fire glared brighter than three suns on a clear day.
So bright that fifty miles away a blind girl turned her head and asked:
"What is that?"
The orb rose, roiling into the sky, sloughing off radioactive skirts of purple, blue, scarlet, green.

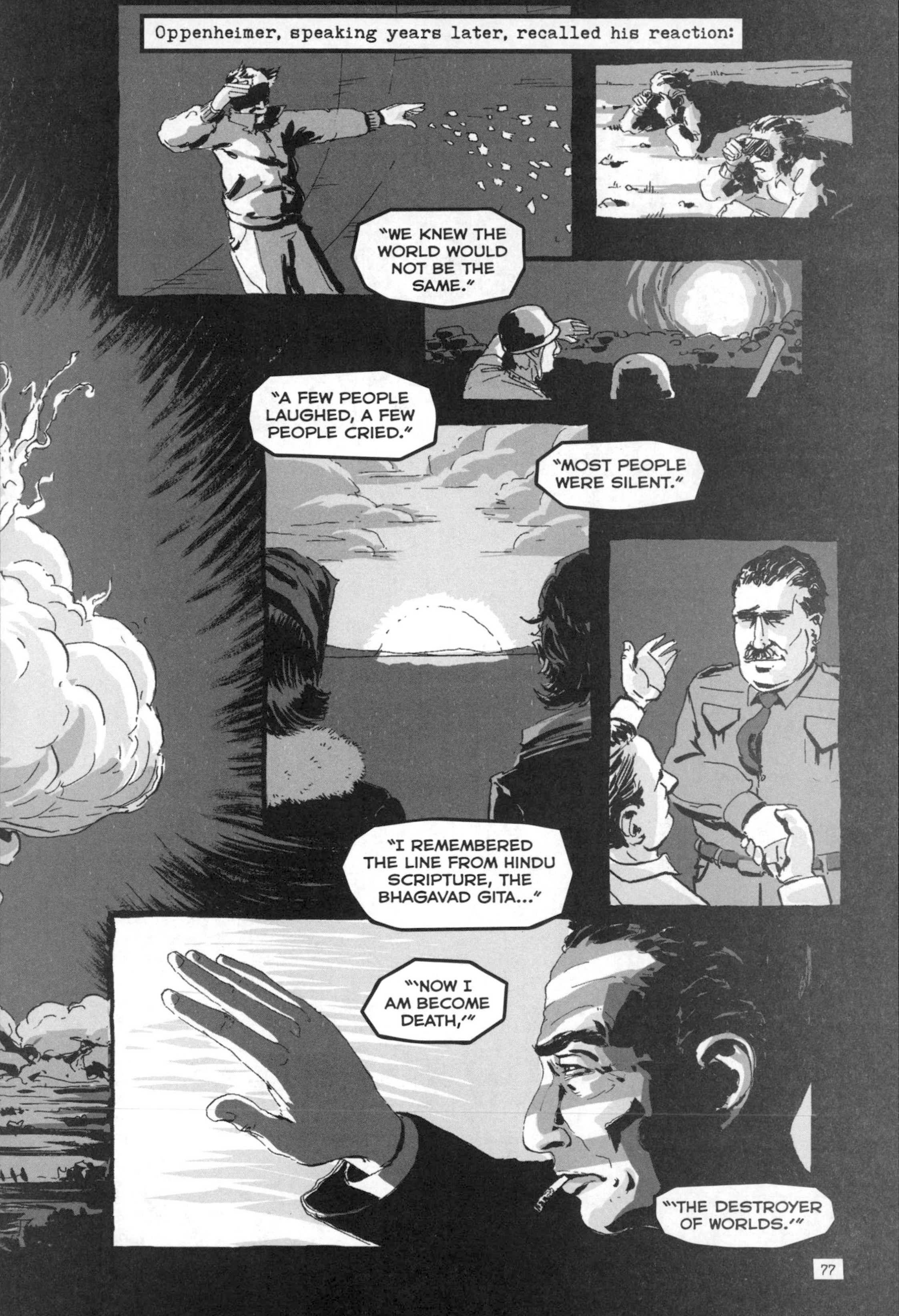

Oppenheimer, speaking years later, recalled his reaction:
"WE KNEW THE WORLD WOULD NOT BE THE SAME."
"A FEW PEOPLE LAUGHED, A FEW PEOPLE CRIED."
"MOST PEOPLE WERE SILENT."
"I REMEMBERED THE LINE FROM HINDU SCRIPTURE, THE BHAGAVAD GITA..."
"'NOW I AM BECOME DEATH,'"
"THE DESTROYER OF WORLDS.'"

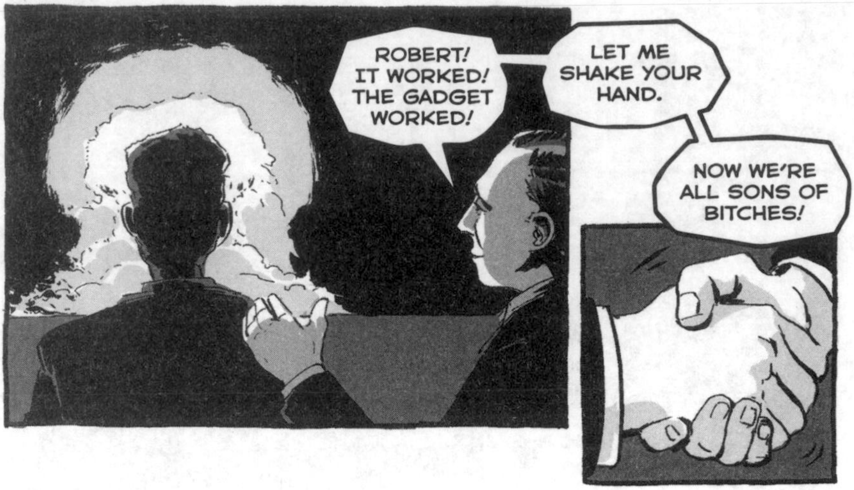
ROBERT! IT WORKED! THE GADGET WORKED!
LET ME SHAKE YOUR HAND.
NOW WE'RE ALL SONS OF BITCHES!

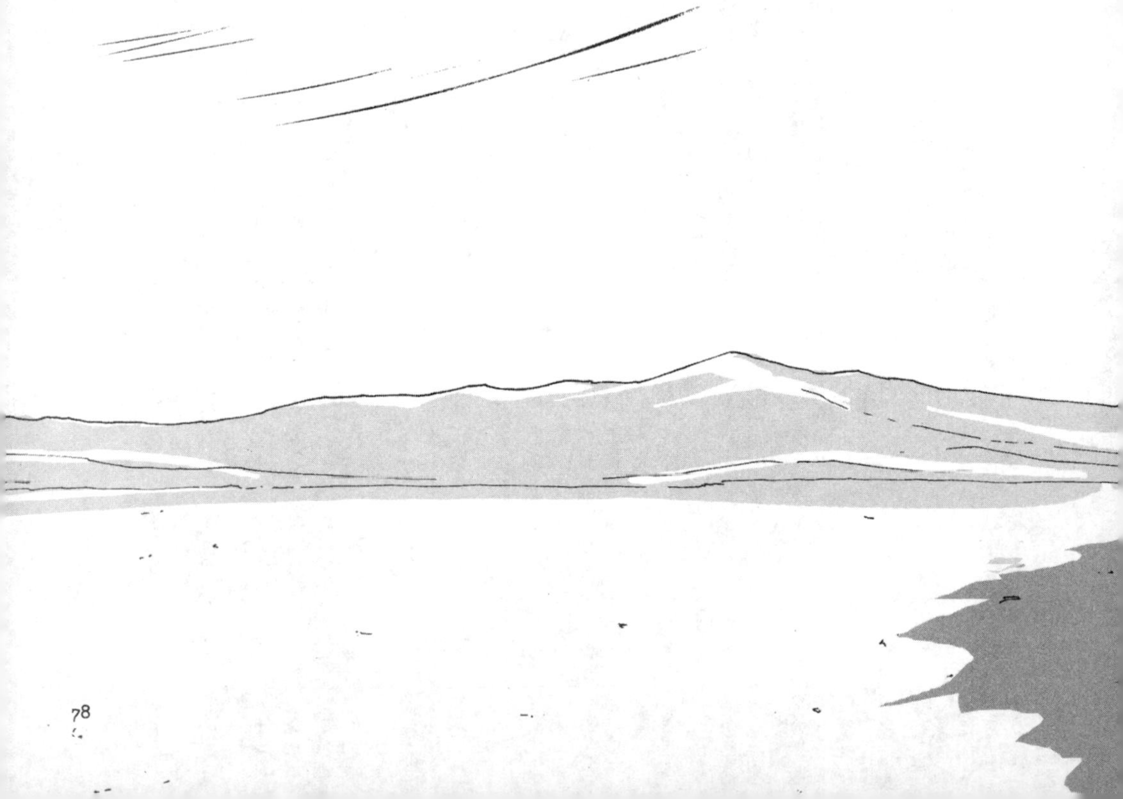

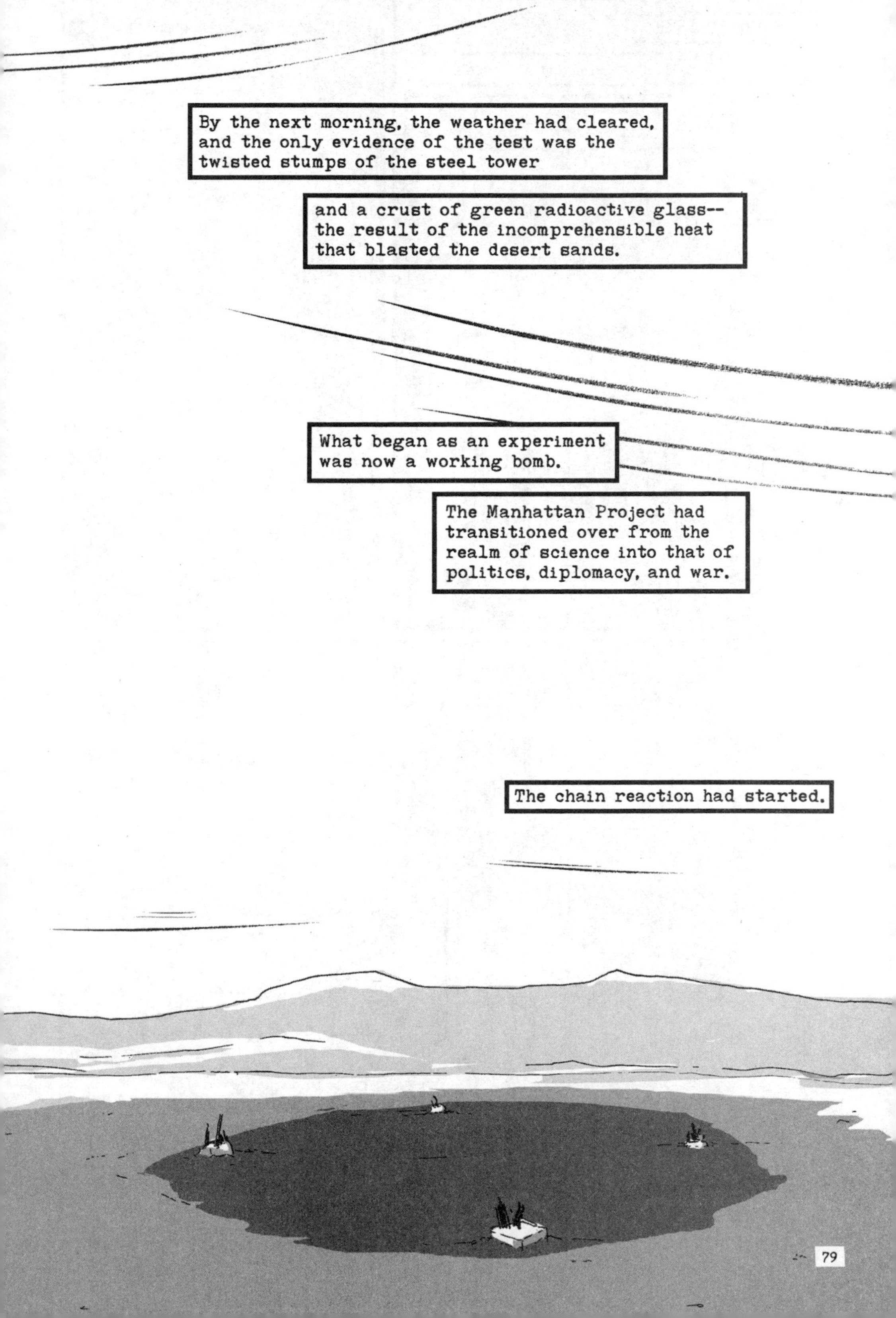
By the next morning, the weather had cleared, and the only evidence of the test was the twisted stumps of the steel tower
and a crust of green radioactive glass--the result of the incomprehensible heat that blasted the desert sands.
What began as an experiment was now a working bomb.
The Manhattan Project had transitioned over from the realm of science into that of politics, diplomacy, and war.
The chain reaction had started.

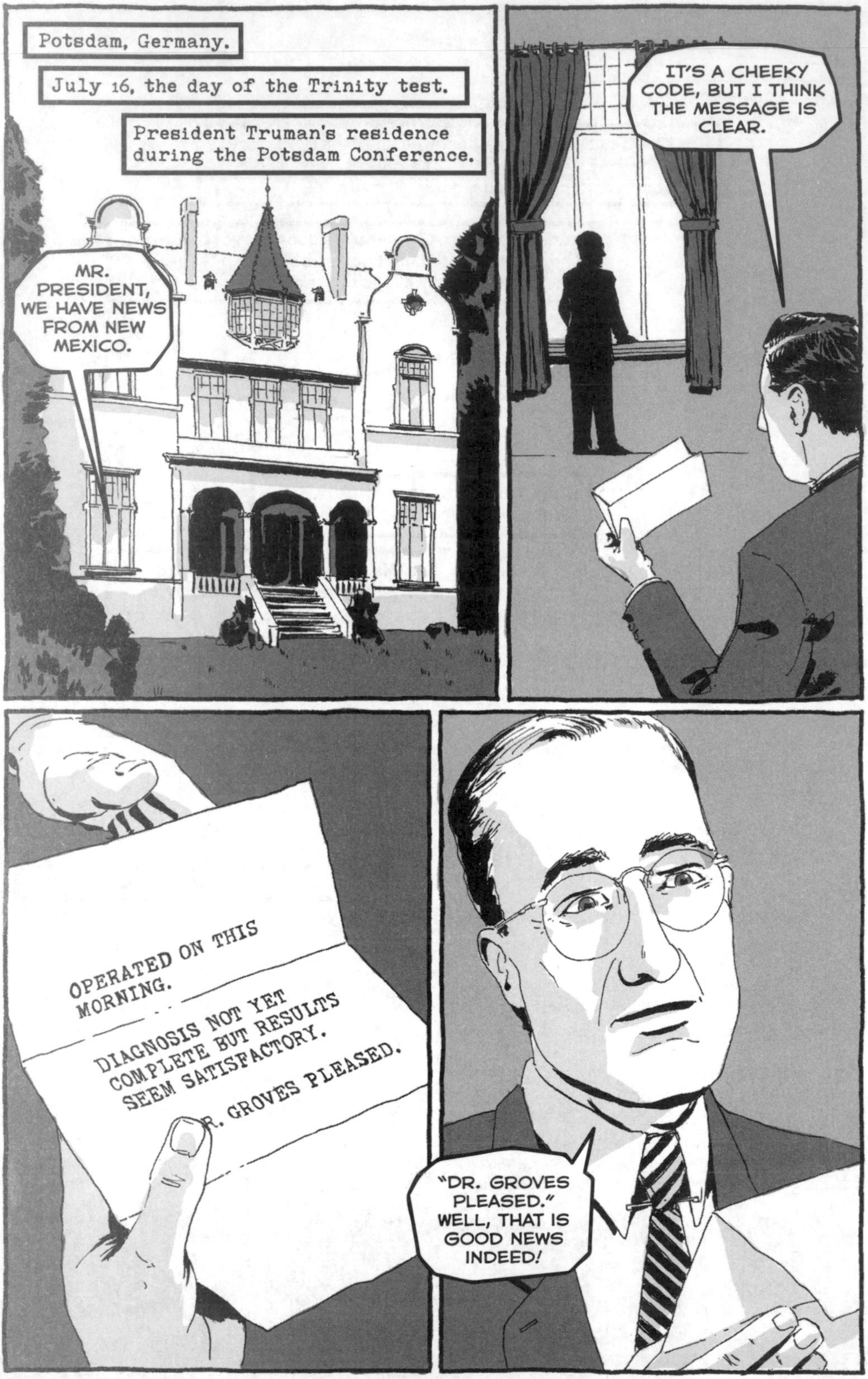
Potsdam, Germany.
July 16, the day of the Trinity test.
President Truman's residence during the Potsdam Conference.
MR. PRESIDENT, WE HAVE NEWS FROM NEW MEXICO.
IT'S A CHEEKY CODE, BUT I THINK THE MESSAGE IS CLEAR.
OPERATED ON THIS MORNING.
DIAGNOSIS NOT YET COMPLETE BUT RESULTS SEEM SATISFACTORY.
R. GROVES PLEASED.
"DR. GROVES PLEASED." WELL, THAT IS GOOD NEWS INDEED!

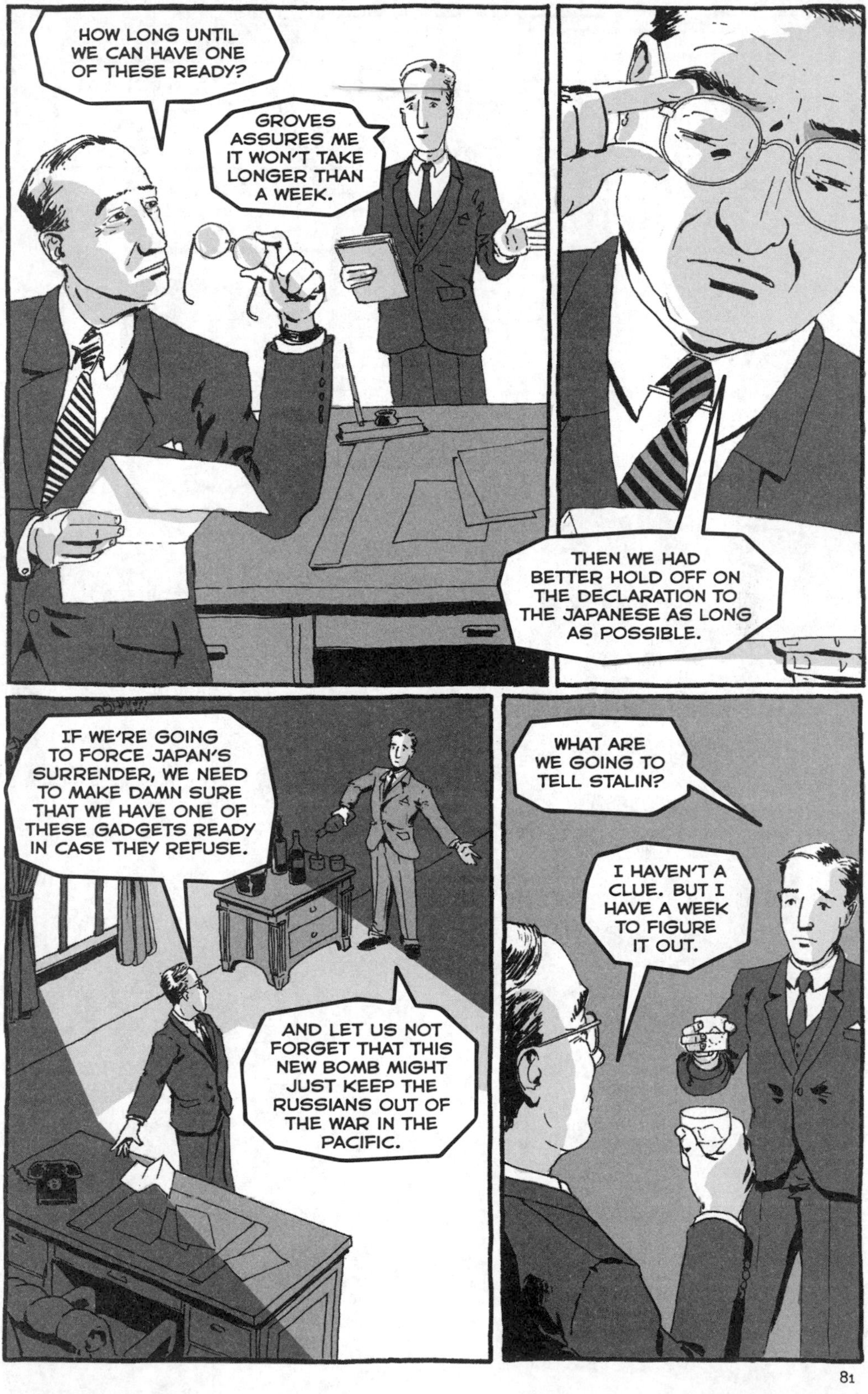
HOW LONG UNTIL WE CAN HAVE ONE OF THESE READY?
GROVES ASSURES ME IT WON'T TAKE LONGER THAN A WEEK.
THEN WE HAD BETTER HOLD OFF ON THE DECLARATION TO THE JAPANESE AS LONG AS POSSIBLE.
IF WE'RE GOING TO FORCE JAPAN'S SURRENDER, WE NEED TO MAKE DAMN SURE THAT WE HAVE ONE OF THESE GADGETS READY IN CASE THEY REFUSE.
AND LET US NOT FORGET THAT THIS NEW BOMB MIGHT JUST KEEP THE RUSSIANS OUT OF THE WAR IN THE PACIFIC.
WHAT ARE WE GOING TO TELL STALIN?
I HAVEN'T A CLUE. BUT I HAVE A WEEK TO FIGURE IT OUT.

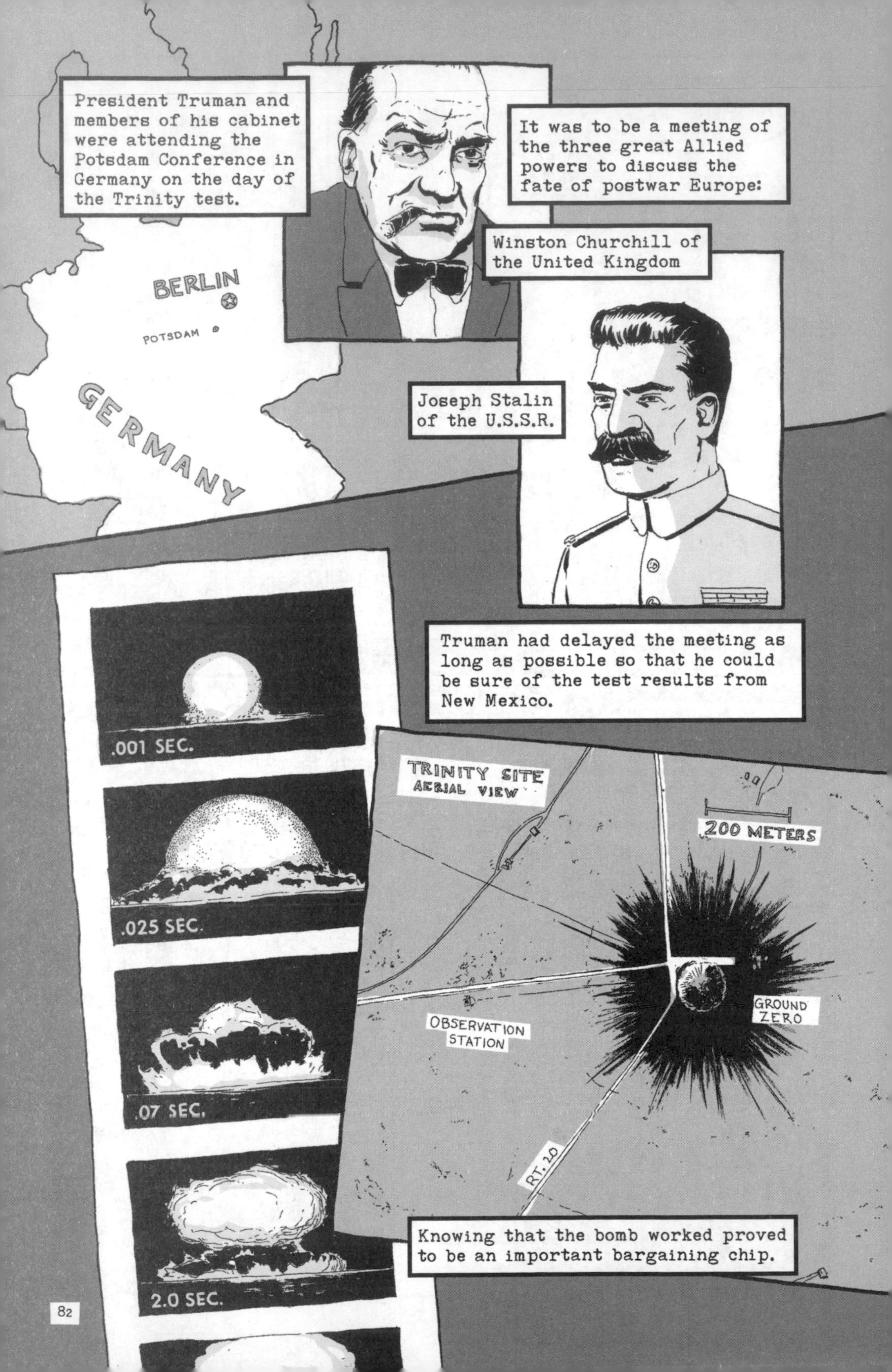

President Truman and members of his cabinet were attending the Potsdam Conference in Germany on the day of the Trinity test.
It was to be a meeting of the three great Allied powers to discuss the fate of postwar Europe:
BERLIN
POTSDAM
GERMANY
Winston Churchill of the United Kingdom
Joseph Stalin of the U.S.S.R.
Truman had delayed the meeting as long as possible so that he could be sure of the test results from New Mexico.
.001 SEC.
.025 SEC.
.07 SEC.
2.0 SEC.
TRINITY SITE AERIAL VIEW
200 METERS
OBSERVATION STATION
GROUND ZERO
RT. 20
Knowing that the bomb worked proved to be an important bargaining chip.

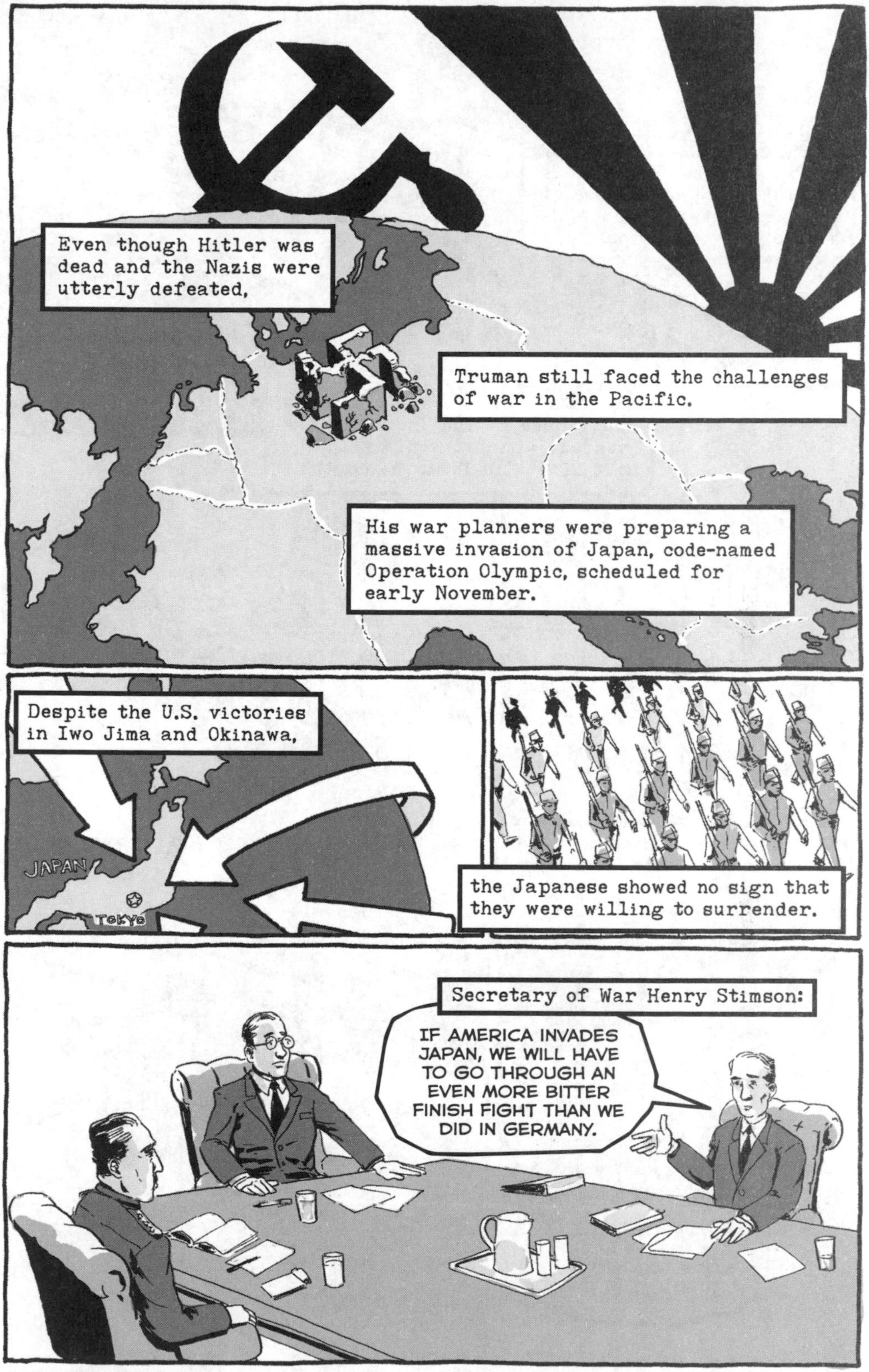
Even though Hitler was dead and the Nazis were utterly defeated,
Truman still faced the challenges of war in the Pacific.
His war planners were preparing a massive invasion of Japan, code-named Operation Olympic, scheduled for early November.
Despite the U.S. victories in Iwo Jima and Okinawa,
JAPAN
TOKYO
the Japanese showed no sign that they were willing to surrender.
Secretary of War Henry Stimson:
IF AMERICA INVADES JAPAN, WE WILL HAVE TO GO THROUGH AN EVEN MORE BITTER FINISH FIGHT THAN WE DID IN GERMANY.

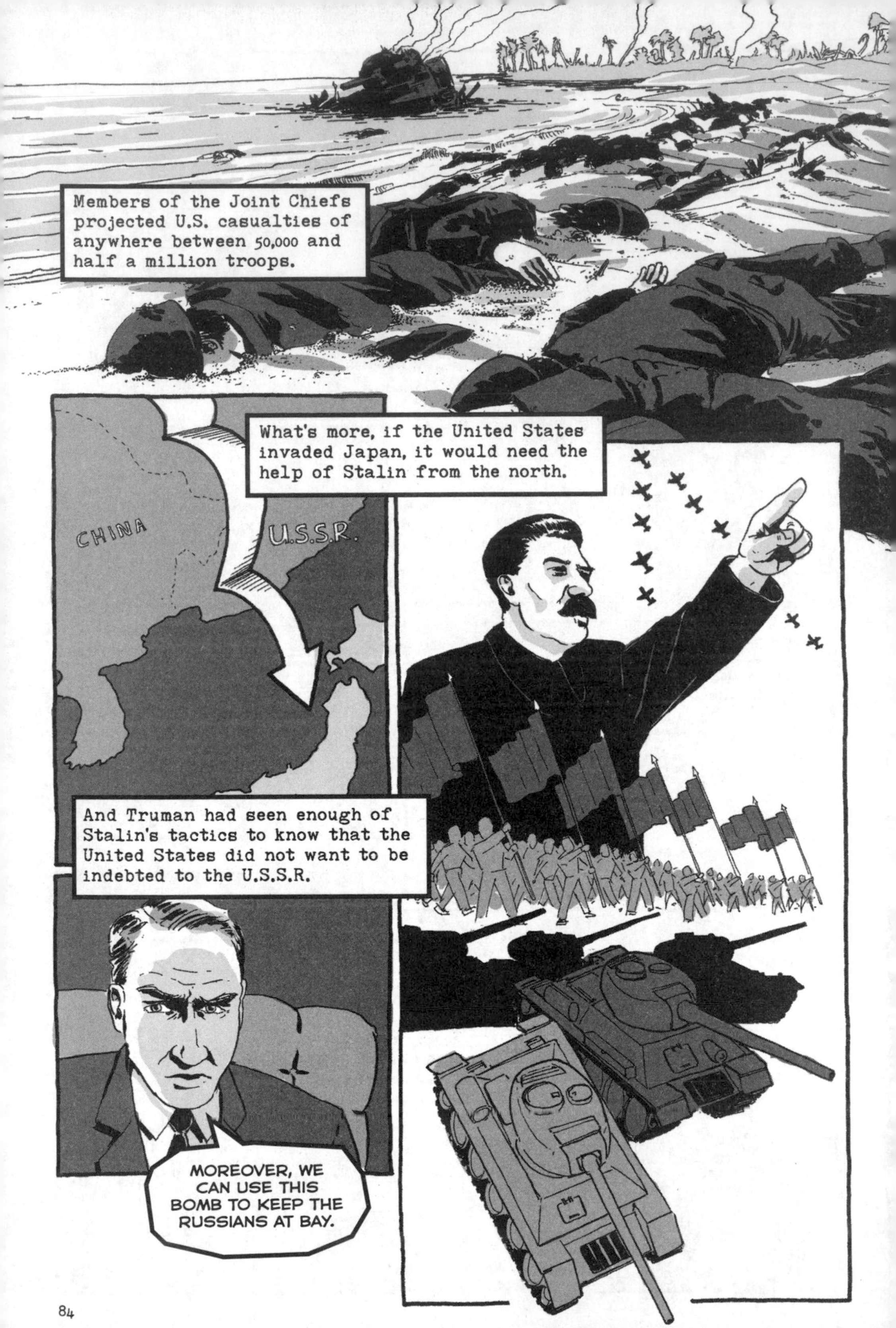
Members of the Joint Chiefs projected U.S. casualties of anywhere between 50,000 and half a million troops.
What's more, if the United States invaded Japan, it would need the help of Stalin from the north.
CHINA
U.S.S.R.
And Truman had seen enough of Stalin's tactics to know that the United States did not want to be indebted to the U.S.S.R.
MOREOVER, WE CAN USE THIS BOMB TO KEEP THE RUSSIANS AT BAY.

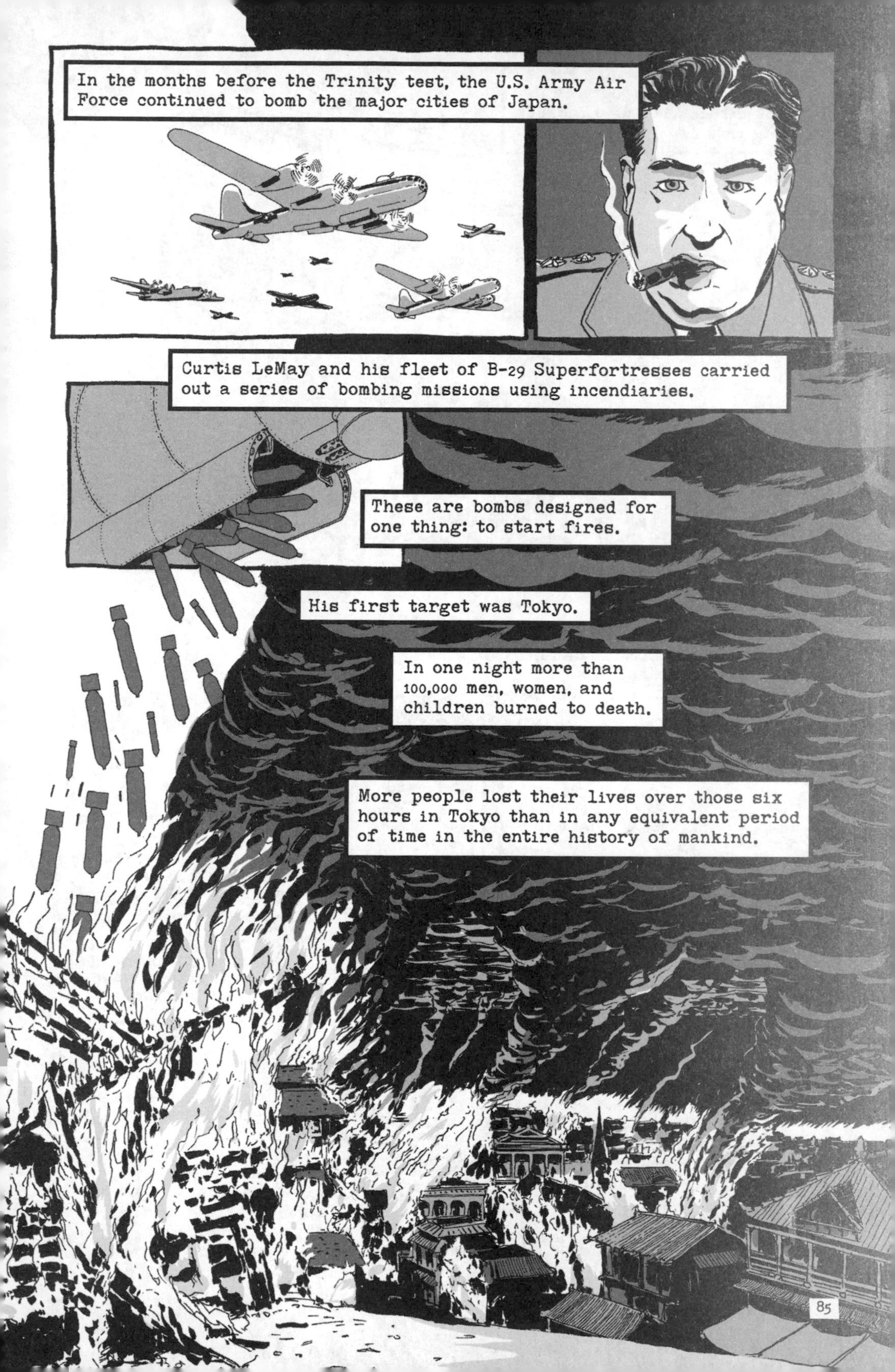

In the months before the Trinity test, the U.S. Army Air Force continued to bomb the major cities of Japan.
Curtis LeMay and his fleet of B-29 Superfortresses carried out a series of bombing missions using incendiaries.
These are bombs designed for one thing: to start fires.
His first target was Tokyo.
In one night more than 100,000 men, women, and children burned to death.
More people lost their lives over those six hours in Tokyo than in any equivalent period of time in the entire history of mankind.

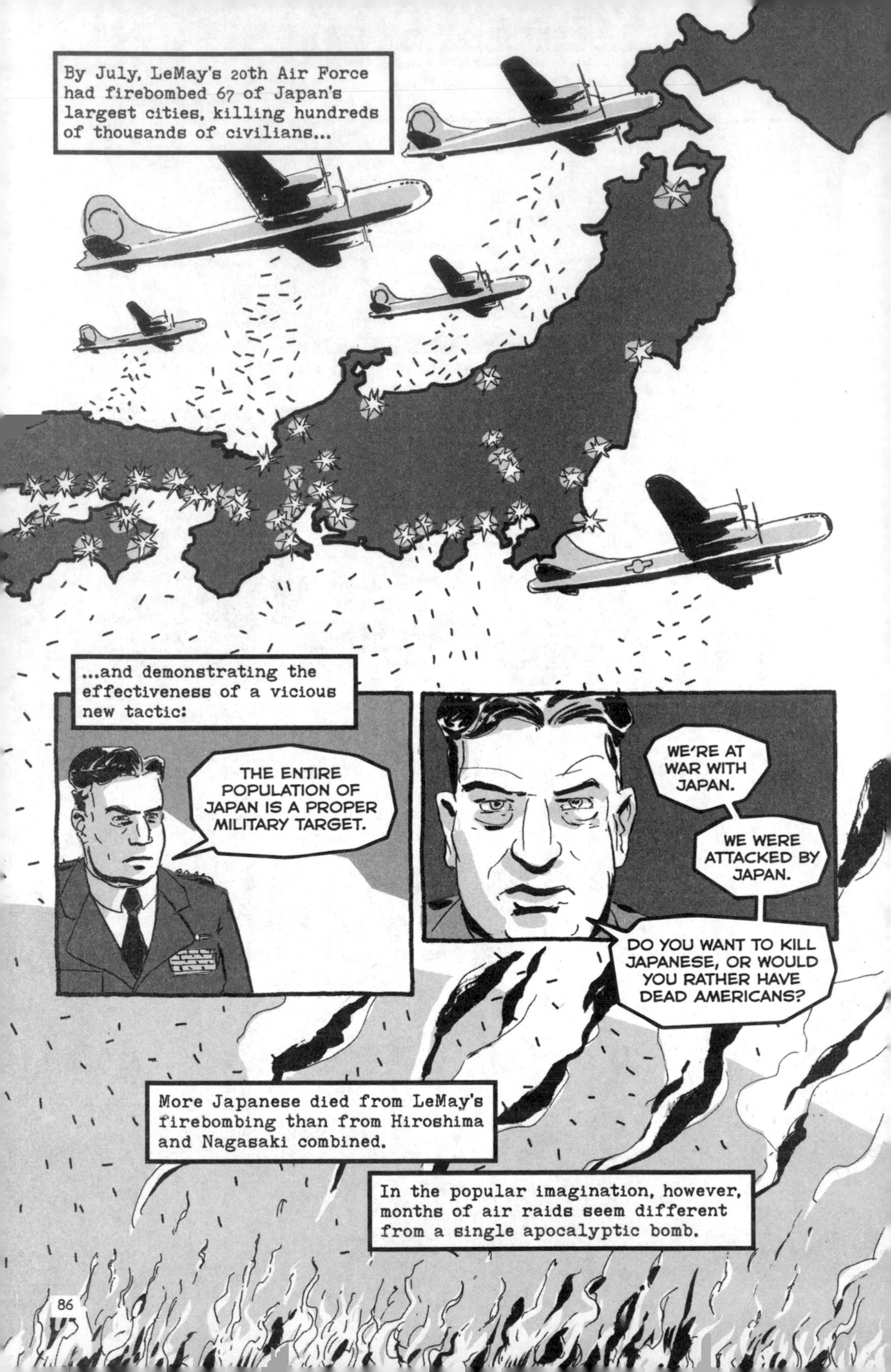
By July, LeMay's 20th Air Force had firebombed 67 of Japan's largest cities, killing hundreds of thousands of civilians...
...and demonstrating the effectiveness of a vicious new tactic:
THE ENTIRE POPULATION OF JAPAN IS A PROPER MILITARY TARGET.
WE'RE AT WAR WITH JAPAN.
WE WERE ATTACKED BY JAPAN.
DO YOU WANT TO KILL JAPANESE, OR WOULD YOU RATHER HAVE DEAD AMERICANS?
More Japanese died from LeMay's firebombing than from Hiroshima and Nagasaki combined.
In the popular imagination, however, months of air raids seem different from a single apocalyptic bomb.

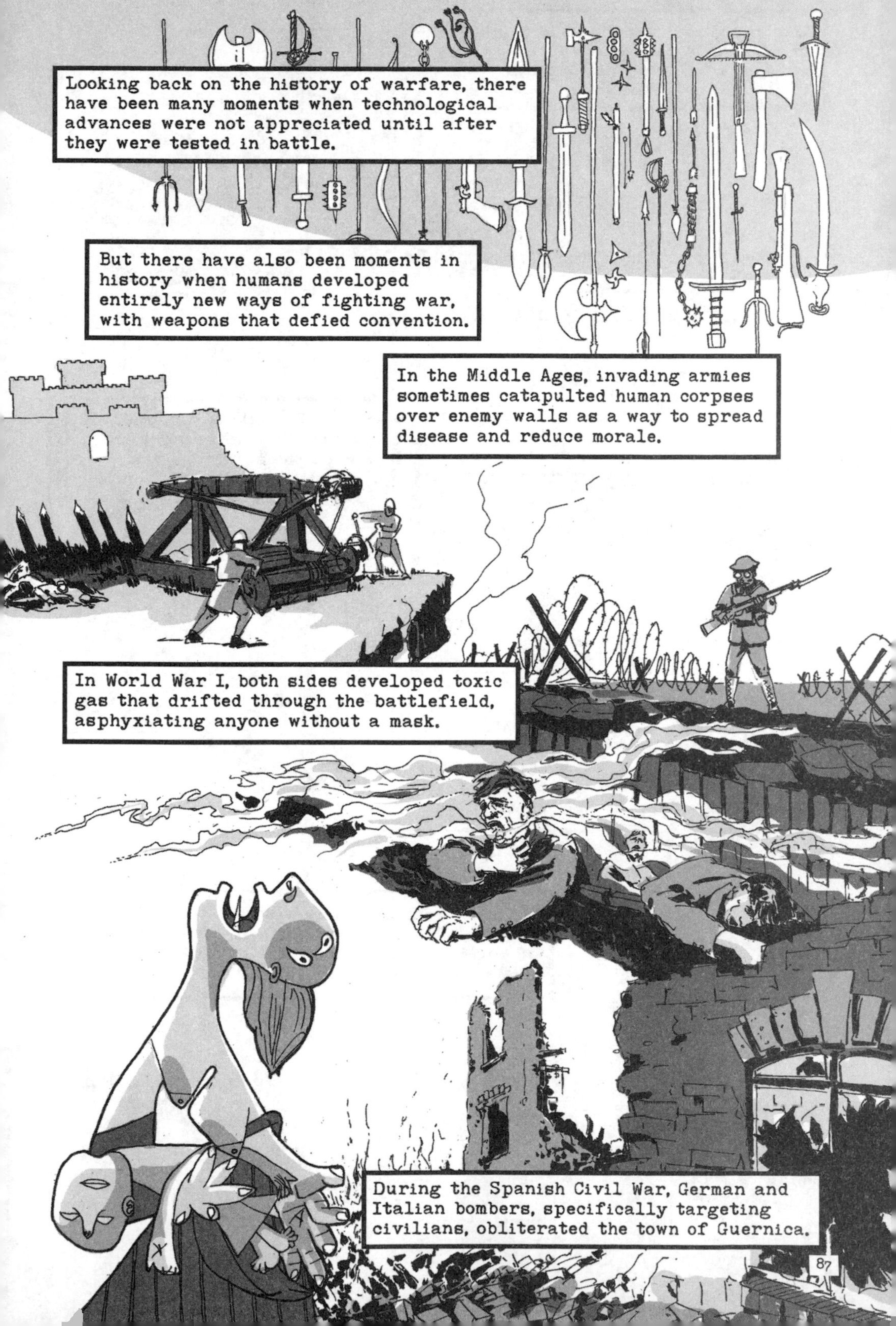
Looking back on the history of warfare, there have been many moments when technological advances were not appreciated until after they were tested in battle.
But there have also been moments in history when humans developed entirely new ways of fighting war, with weapons that defied convention.
In the Middle Ages, invading armies sometimes catapulted human corpses over enemy walls as a way to spread disease and reduce morale.
In World War I, both sides developed toxic gas that drifted through the battlefield, asphyxiating anyone without a mask.
During the Spanish Civil War, German and Italian bombers, specifically targeting civilians, obliterated the town of Guernica.

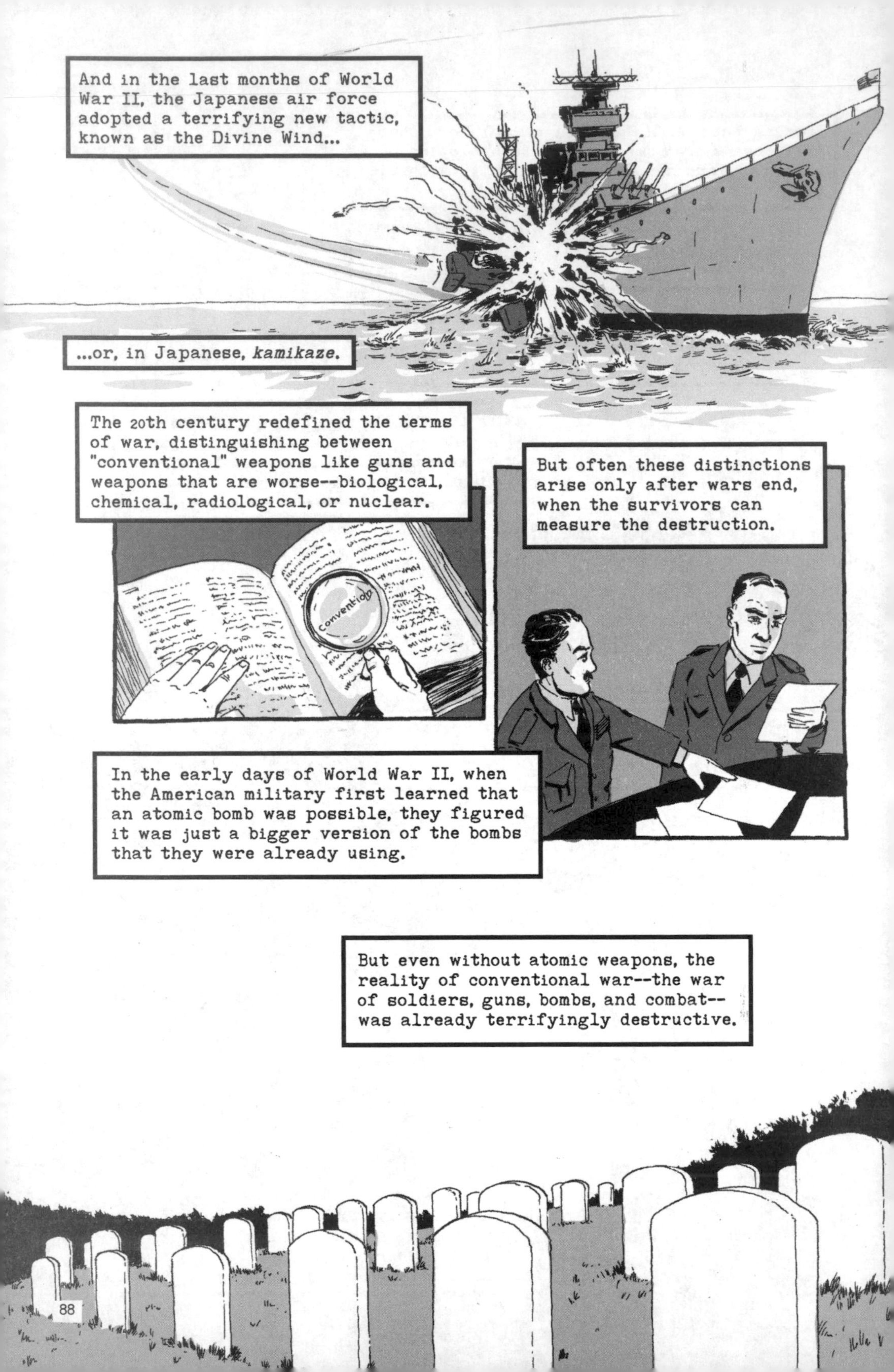
And in the last months of World War II, the Japanese air force adopted a terrifying new tactic, known as the Divine Wind...
...or, in Japanese, *kamikaze*.
The 20th century redefined the terms of war, distinguishing between "conventional" weapons like guns and weapons that are worse--biological, chemical, radiological, or nuclear.
But often these distinctions arise only after wars end, when the survivors can measure the destruction.
In the early days of World War II, when the American military first learned that an atomic bomb was possible, they figured it was just a bigger version of the bombs that they were already using.
But even without atomic weapons, the reality of conventional war--the war of soldiers, guns, bombs, and combat--was already terrifyingly destructive.

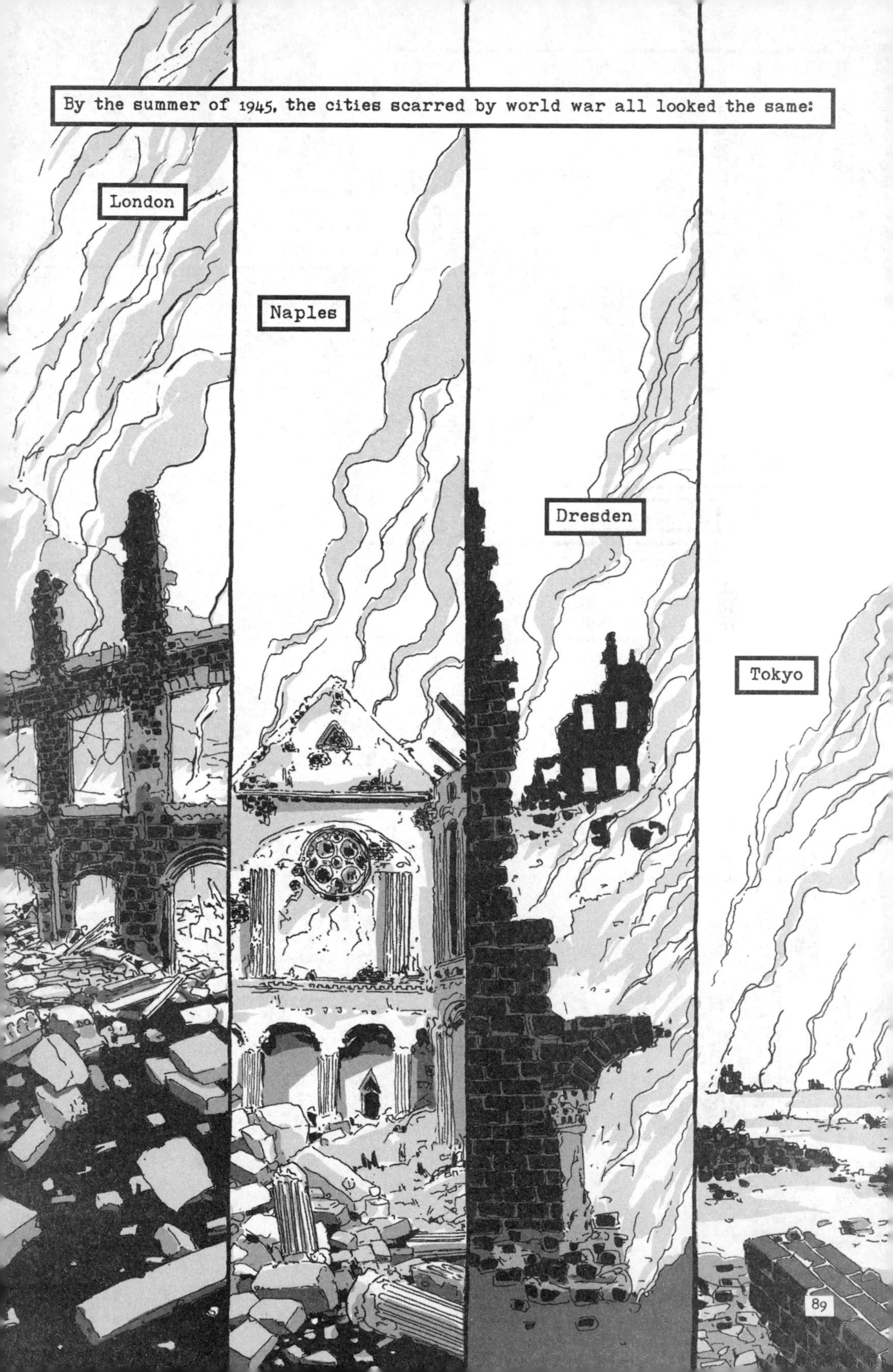

By the summer of 1945, the cities scarred by world war all looked the same:
London
Naples
Dresden
Tokyo

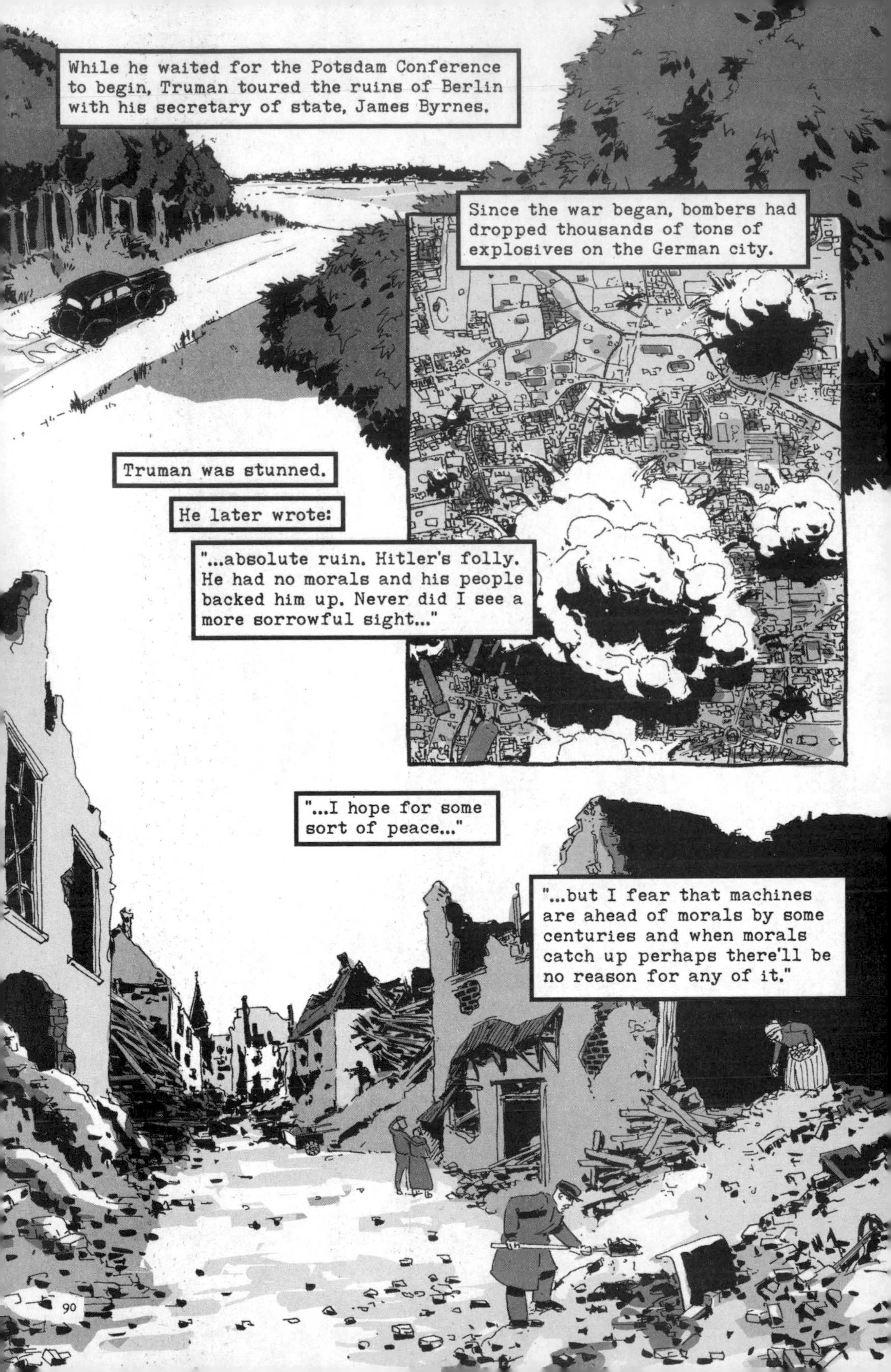
While he waited for the Potsdam Conference to begin, Truman toured the ruins of Berlin with his secretary of state, James Byrnes.
Since the war began, bombers had dropped thousands of tons of explosives on the German city.
Truman was stunned.
He later wrote:
"...absolute ruin. Hitler's folly. He had no morals and his people backed him up. Never did I see a more sorrowful sight..."
"...I hope for some sort of peace..."
"...but I fear that machines are ahead of morals by some centuries and when morals catch up perhaps there'll be no reason for any of it."

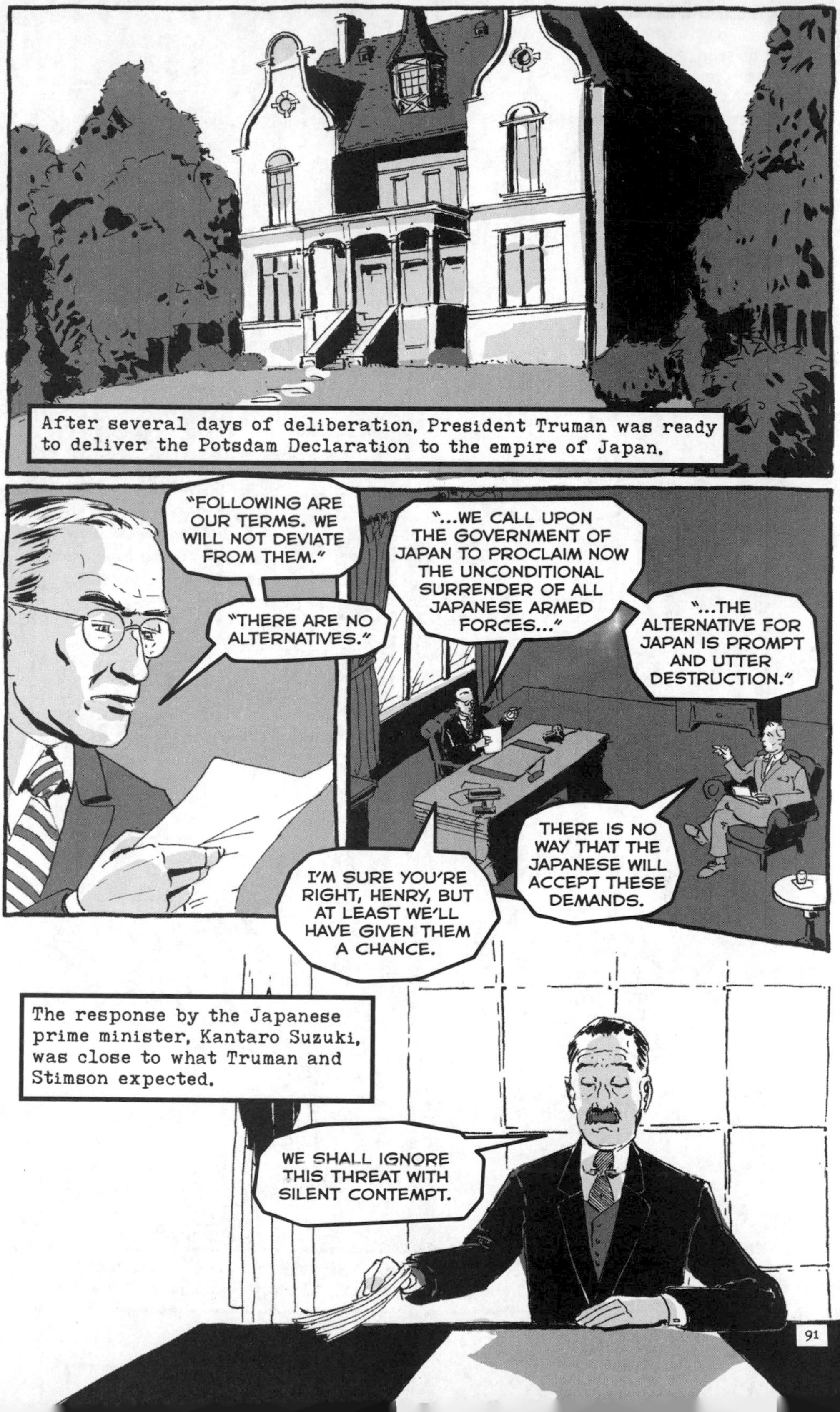
After several days of deliberation, President Truman was ready to deliver the Potsdam Declaration to the empire of Japan.
"FOLLOWING ARE OUR TERMS. WE WILL NOT DEVIATE FROM THEM."
"THERE ARE NO ALTERNATIVES."
"...WE CALL UPON THE GOVERNMENT OF JAPAN TO PROCLAIM NOW THE UNCONDITIONAL SURRENDER OF ALL JAPANESE ARMED FORCES..."
"...THE ALTERNATIVE FOR JAPAN IS PROMPT AND UTTER DESTRUCTION."
THERE IS NO WAY THAT THE JAPANESE WILL ACCEPT THESE DEMANDS.
I'M SURE YOU'RE RIGHT, HENRY, BUT AT LEAST WE'LL HAVE GIVEN THEM A CHANCE.
The response by the Japanese prime minister, Kantaro Suzuki, was close to what Truman and Stimson expected.
WE SHALL IGNORE THIS THREAT WITH SILENT CONTEMPT.

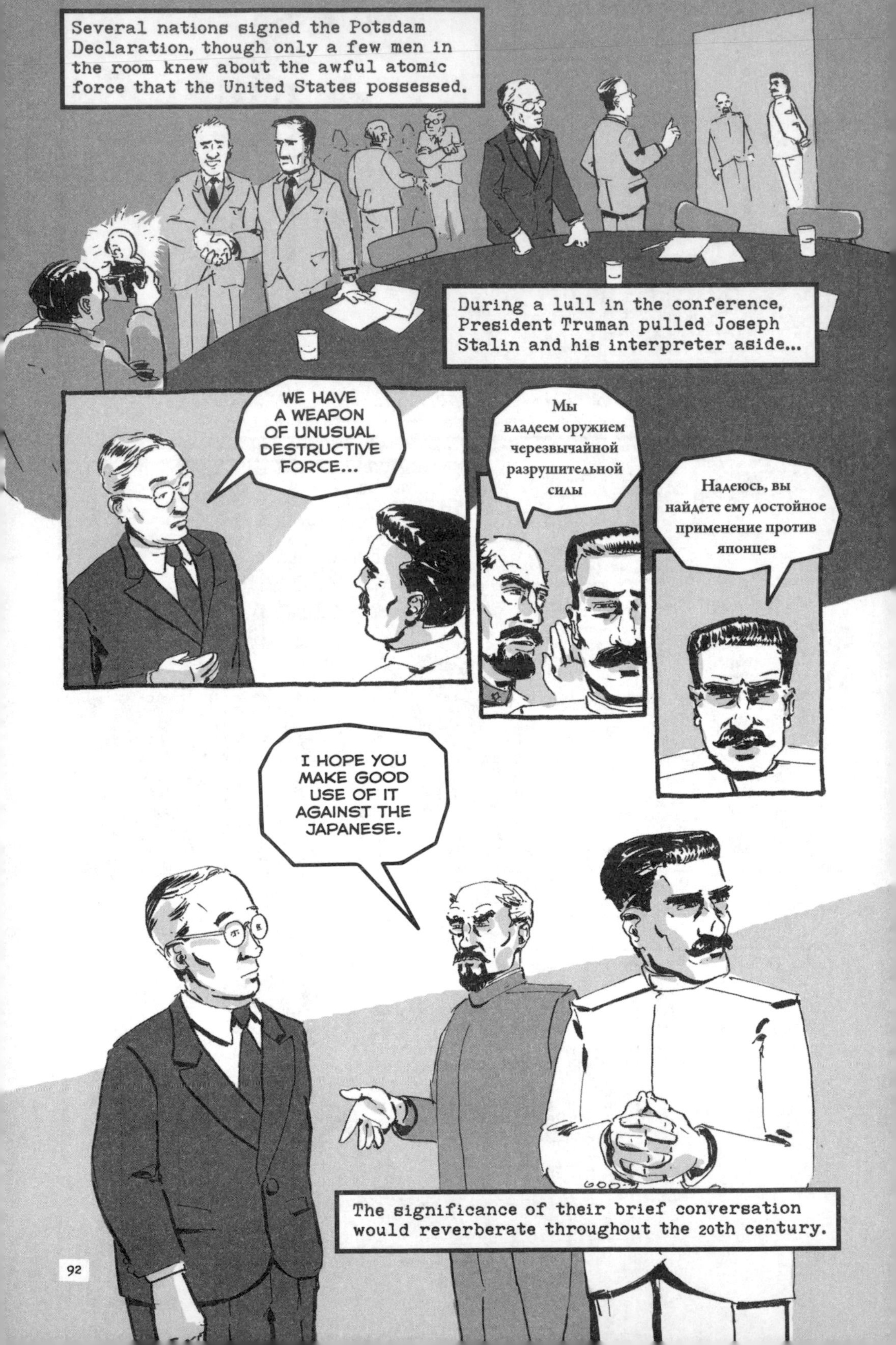
Several nations signed the Potsdam Declaration, though only a few men in the room knew about the awful atomic force that the United States possessed.
During a lull in the conference, President Truman pulled Joseph Stalin and his interpreter aside...
WE HAVE A WEAPON OF UNUSUAL DESTRUCTIVE FORCE...
Мы владеем оружием чрезвычайной разрушительной силы
Надеюсь, вы найдете ему достойное применение против японцев
I HOPE YOU MAKE GOOD USE OF IT AGAINST THE JAPANESE.
The significance of their brief conversation would reverberate throughout the 20th century.

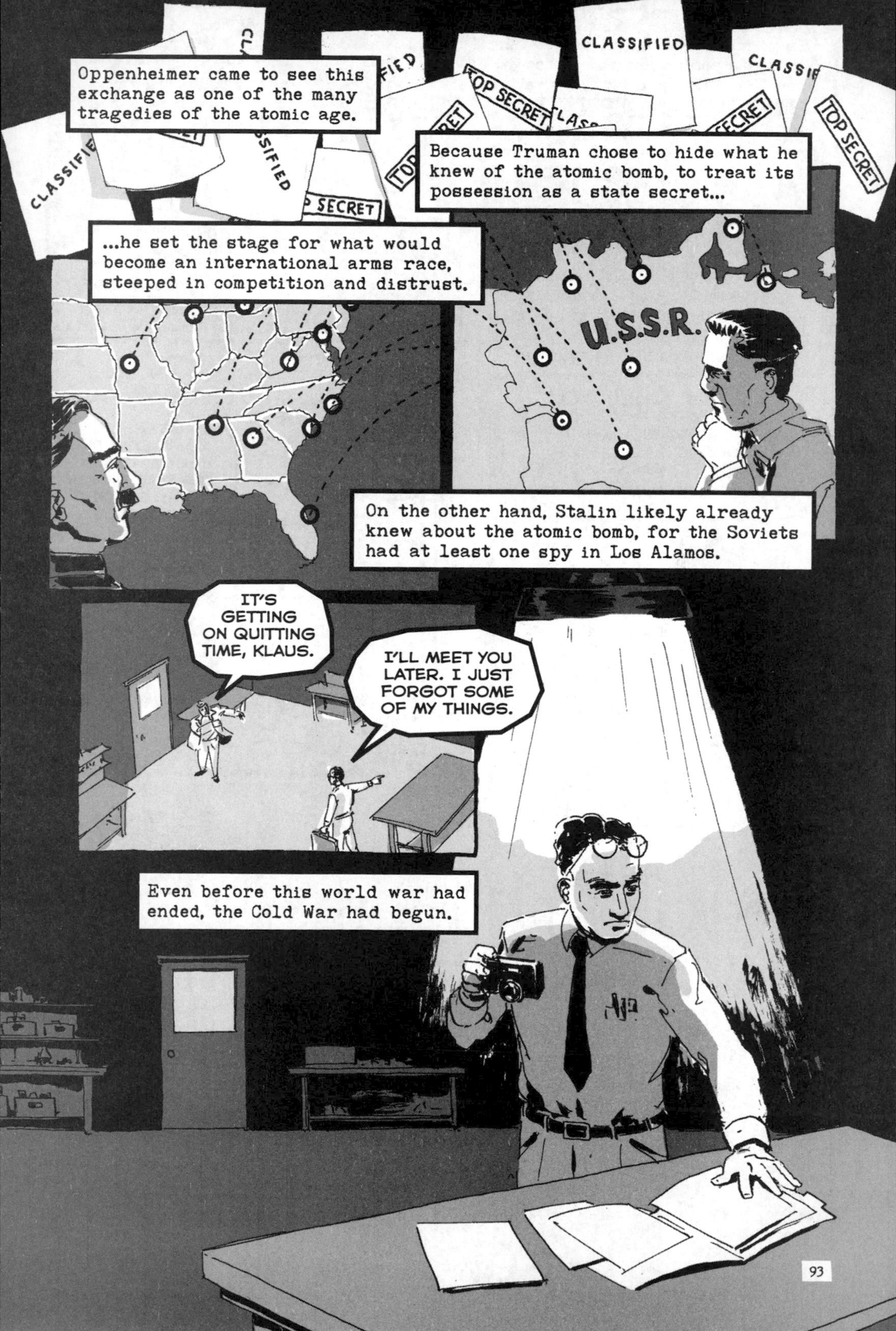
CLASSIFIED
TOP SECRET
Oppenheimer came to see this exchange as one of the many tragedies of the atomic age.
Because Truman chose to hide what he knew of the atomic bomb, to treat its possession as a state secret...
...he set the stage for what would become an international arms race, steeped in competition and distrust.
U.S.S.R.
On the other hand, Stalin likely already knew about the atomic bomb, for the Soviets had at least one spy in Los Alamos.
IT'S GETTING ON QUITTING TIME, KLAUS.
I'LL MEET YOU LATER. I JUST FORGOT SOME OF MY THINGS.
Even before this world war had ended, the Cold War had begun.

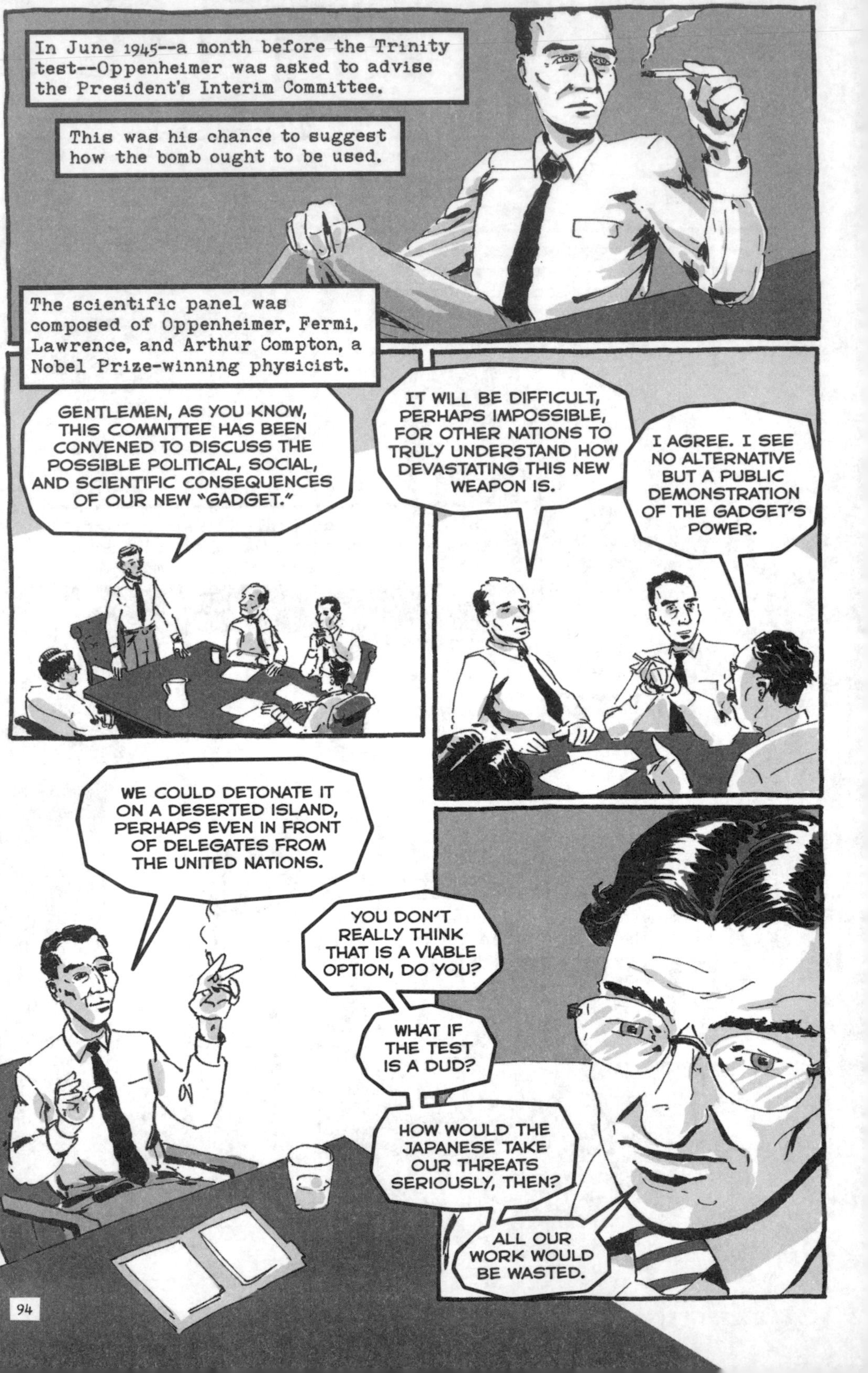
In June 1945--a month before the Trinity test--Oppenheimer was asked to advise the President's Interim Committee.
This was his chance to suggest how the bomb ought to be used.
The scientific panel was composed of Oppenheimer, Fermi, Lawrence, and Arthur Compton, a Nobel Prize-winning physicist.
GENTLEMEN, AS YOU KNOW, THIS COMMITTEE HAS BEEN CONVENED TO DISCUSS THE POSSIBLE POLITICAL, SOCIAL, AND SCIENTIFIC CONSEQUENCES OF OUR NEW "GADGET."
IT WILL BE DIFFICULT, PERHAPS IMPOSSIBLE, FOR OTHER NATIONS TO TRULY UNDERSTAND HOW DEVASTATING THIS NEW WEAPON IS.
I AGREE. I SEE NO ALTERNATIVE BUT A PUBLIC DEMONSTRATION OF THE GADGET'S POWER.
WE COULD DETONATE IT ON A DESERTED ISLAND, PERHAPS EVEN IN FRONT OF DELEGATES FROM THE UNITED NATIONS.
YOU DON'T REALLY THINK THAT IS A VIABLE OPTION, DO YOU?
WHAT IF THE TEST IS A DUD?
HOW WOULD THE JAPANESE TAKE OUR THREATS SERIOUSLY, THEN?
ALL OUR WORK WOULD BE WASTED.

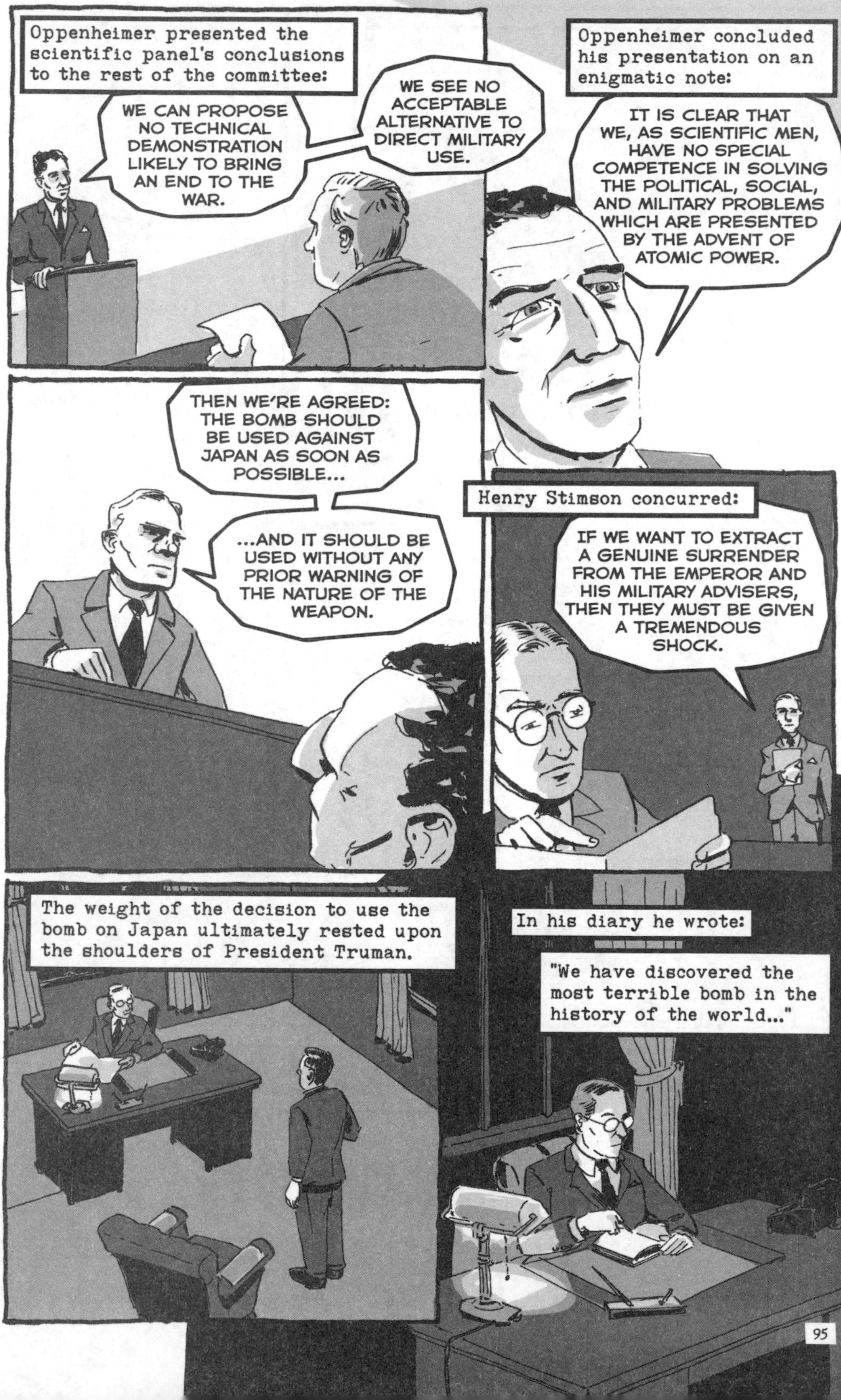
Oppenheimer presented the scientific panel's conclusions to the rest of the committee:
WE CAN PROPOSE NO TECHNICAL DEMONSTRATION LIKELY TO BRING AN END TO THE WAR.
WE SEE NO ACCEPTABLE ALTERNATIVE TO DIRECT MILITARY USE.
Oppenheimer concluded his presentation on an enigmatic note:
IT IS CLEAR THAT WE, AS SCIENTIFIC MEN, HAVE NO SPECIAL COMPETENCE IN SOLVING THE POLITICAL, SOCIAL, AND MILITARY PROBLEMS WHICH ARE PRESENTED BY THE ADVENT OF ATOMIC POWER.
THEN WE'RE AGREED: THE BOMB SHOULD BE USED AGAINST JAPAN AS SOON AS POSSIBLE...
...AND IT SHOULD BE USED WITHOUT ANY PRIOR WARNING OF THE NATURE OF THE WEAPON.
Henry Stimson concurred:
IF WE WANT TO EXTRACT A GENUINE SURRENDER FROM THE EMPEROR AND HIS MILITARY ADVISERS, THEN THEY MUST BE GIVEN A TREMENDOUS SHOCK.
The weight of the decision to use the bomb on Japan ultimately rested upon the shoulders of President Truman.
In his diary he wrote:
"We have discovered the most terrible bomb in the history of the world..."

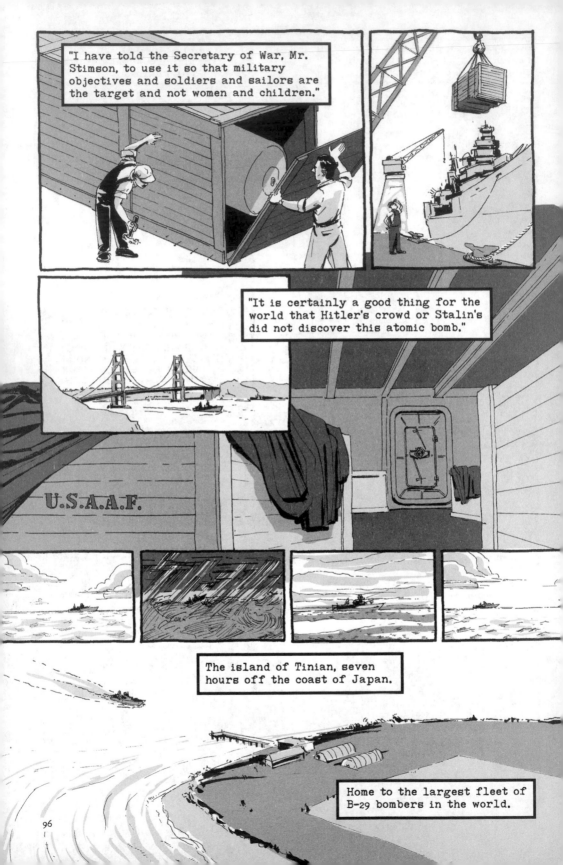

"I have told the Secretary of War, Mr. Stimson, to use it so that military objectives and soldiers and sailors are the target and not women and children."

"It is certainly a good thing for the world that Hitler's crowd or Stalin's did not discover this atomic bomb."

U.S.A.A.F.

The island of Tinian, seven hours off the coast of Japan.

Home to the largest fleet of B-29 bombers in the world.

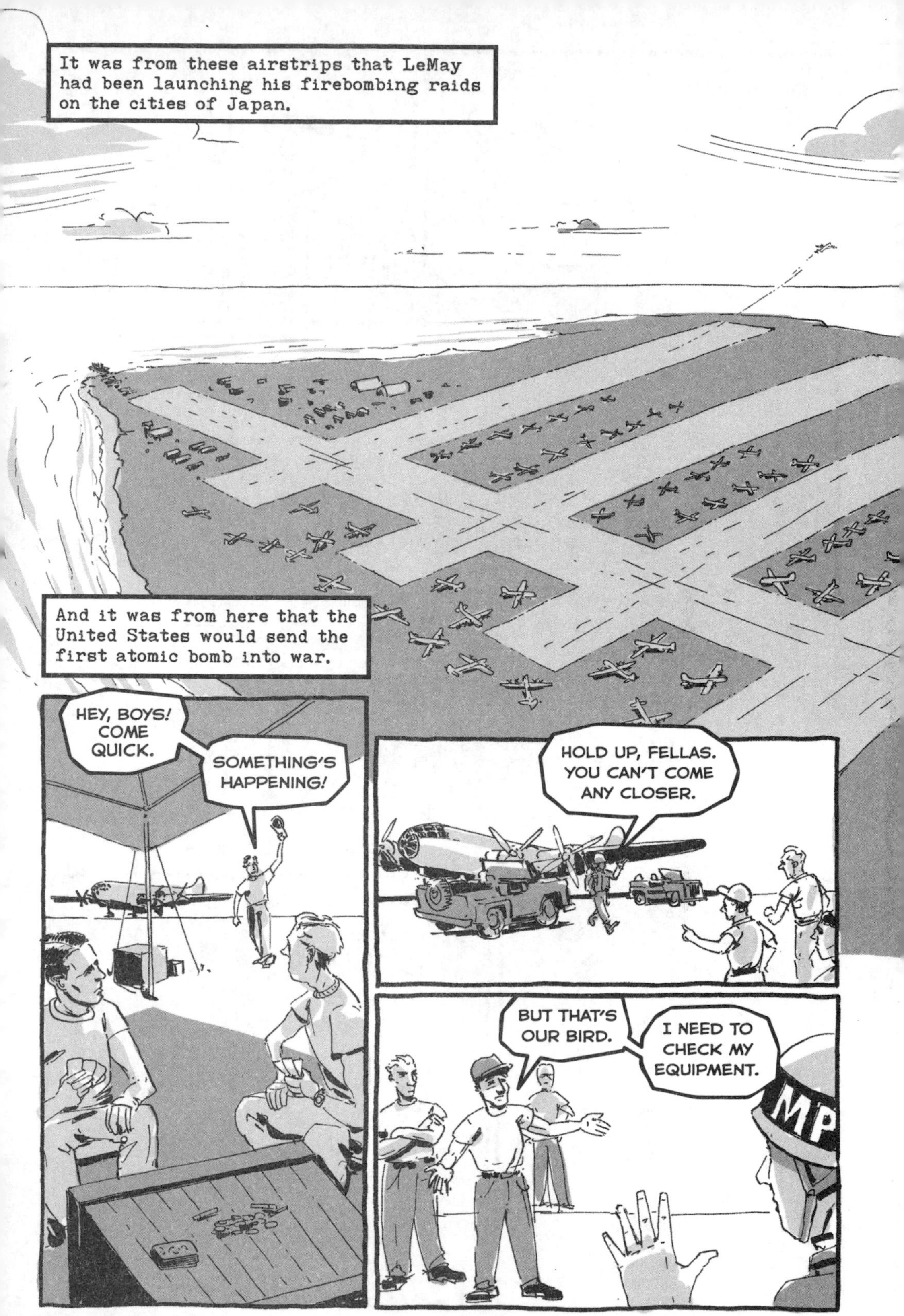
It was from these airstrips that LeMay had been launching his firebombing raids on the cities of Japan.
And it was from here that the United States would send the first atomic bomb into war.
HEY, BOYS! COME QUICK.
SOMETHING'S HAPPENING!
HOLD UP, FELLAS. YOU CAN'T COME ANY CLOSER.
BUT THAT'S OUR BIRD.
I NEED TO CHECK MY EQUIPMENT.
MP

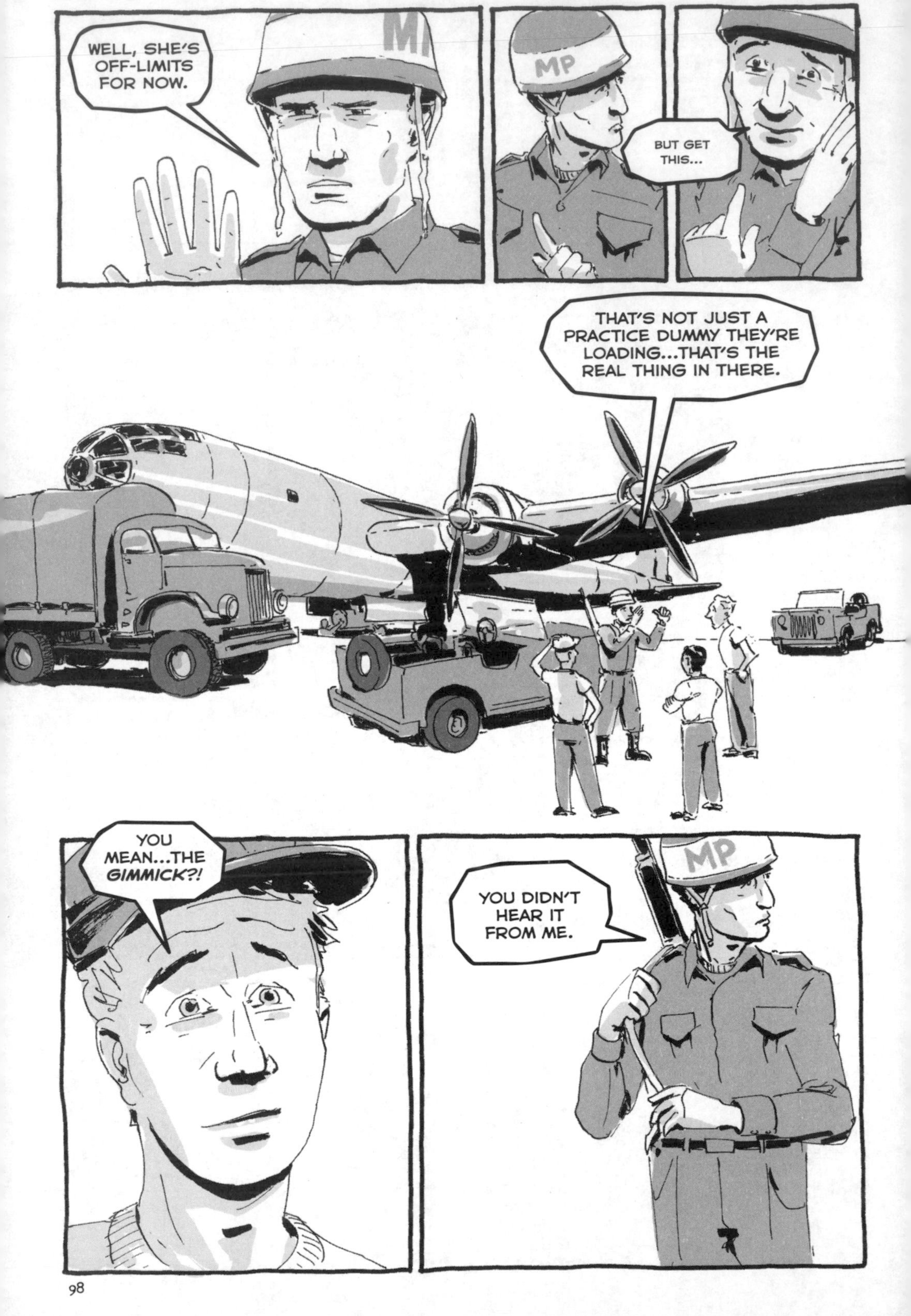
WELL, SHE'S OFF-LIMITS FOR NOW.
MP
BUT GET THIS...
THAT'S NOT JUST A PRACTICE DUMMY THEY'RE LOADING...THAT'S THE REAL THING IN THERE.
YOU MEAN...THE *GIMMICK*?!
MP
YOU DIDN'T HEAR IT FROM ME.

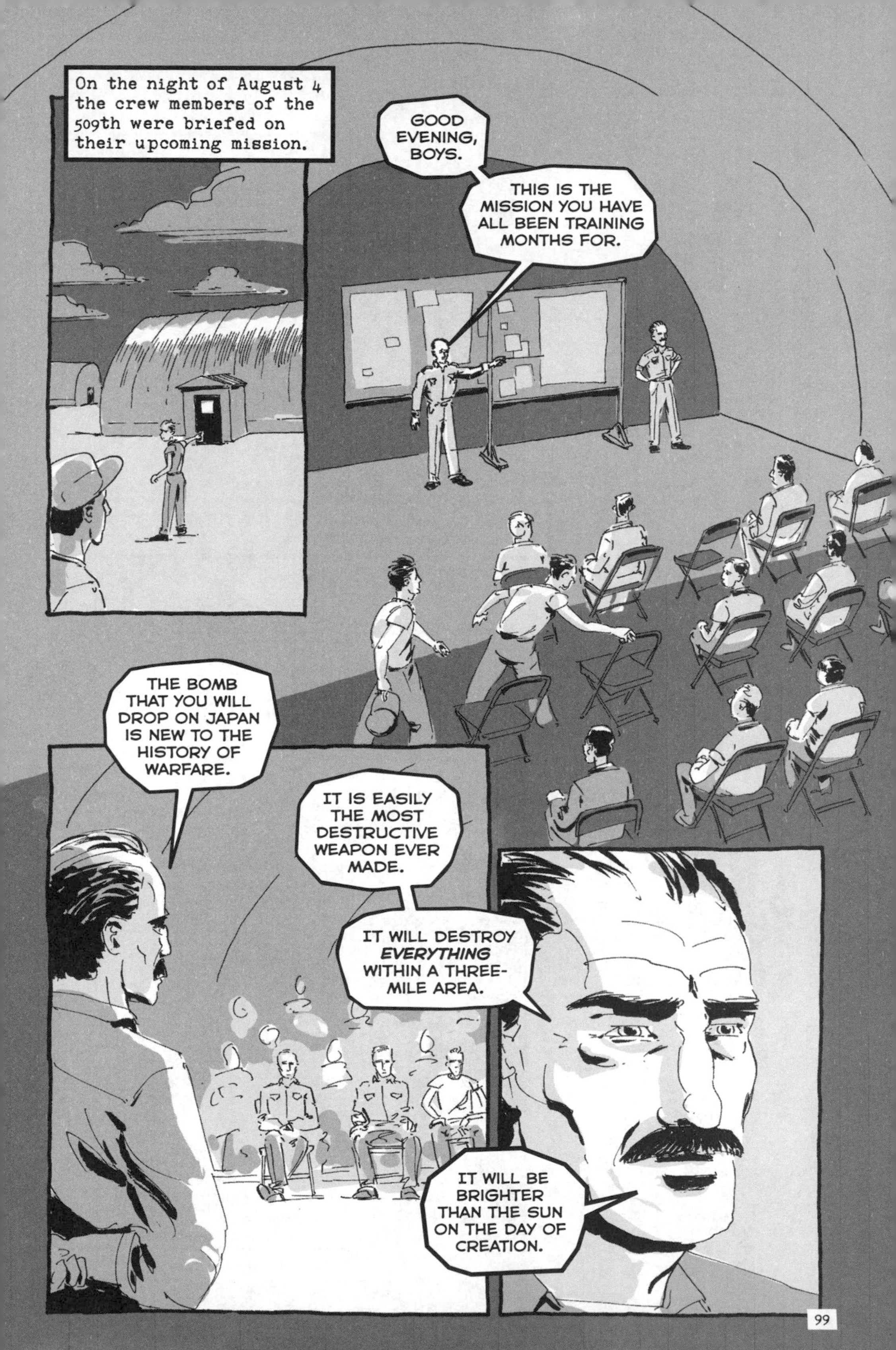
On the night of August 4 the crew members of the 509th were briefed on their upcoming mission.
GOOD EVENING, BOYS.
THIS IS THE MISSION YOU HAVE ALL BEEN TRAINING MONTHS FOR.
THE BOMB THAT YOU WILL DROP ON JAPAN IS NEW TO THE HISTORY OF WARFARE.
IT IS EASILY THE MOST DESTRUCTIVE WEAPON EVER MADE.
IT WILL DESTROY *EVERYTHING* WITHIN A THREE-MILE AREA.
IT WILL BE BRIGHTER THAN THE SUN ON THE DAY OF CREATION.

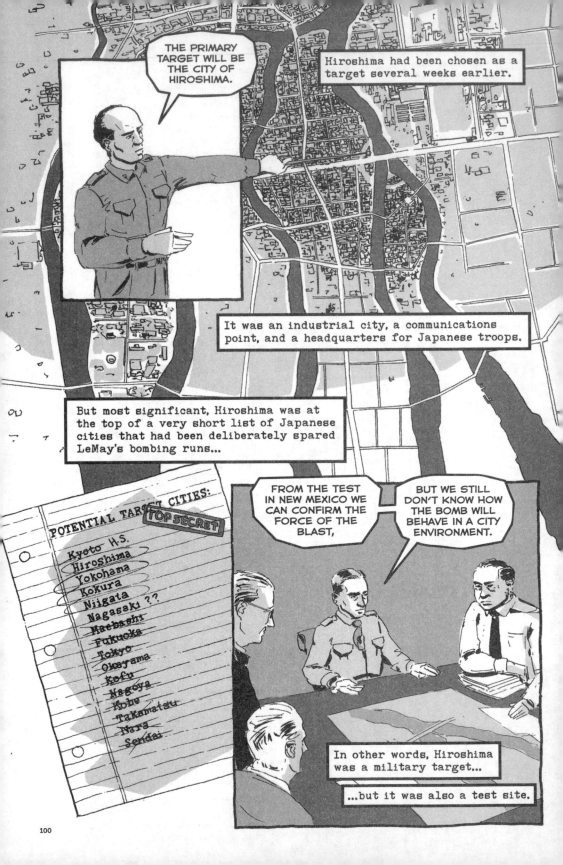

THE PRIMARY TARGET WILL BE THE CITY OF HIROSHIMA.

Hiroshima had been chosen as a target several weeks earlier.

It was an industrial city, a communications point, and a headquarters for Japanese troops.

But most significant, Hiroshima was at the top of a very short list of Japanese cities that had been deliberately spared LeMay's bombing runs...

POTENTIAL TARGET CITIES:
TOP SECRET

Kyoto H.S.
Hiroshima
Yokohama
Kokura
Niigata
Nagasaki ??
Maebashi
Fukuoka
Tokyo
Okayama
Kofu
Nagoya
Kobe
Takamatsu
Nara
Sendai

FROM THE TEST IN NEW MEXICO WE CAN CONFIRM THE FORCE OF THE BLAST,

BUT WE STILL DON'T KNOW HOW THE BOMB WILL BEHAVE IN A CITY ENVIRONMENT.

In other words, Hiroshima was a military target...

...but it was also a test site.

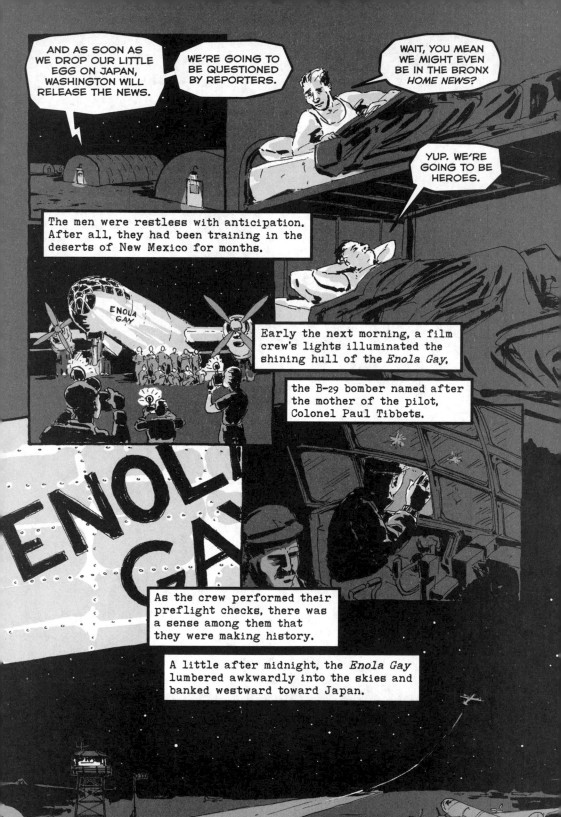

AND AS SOON AS WE DROP OUR LITTLE EGG ON JAPAN, WASHINGTON WILL RELEASE THE NEWS.

WE'RE GOING TO BE QUESTIONED BY REPORTERS.

WAIT, YOU MEAN WE MIGHT EVEN BE IN THE BRONX *HOME NEWS?*

YUP. WE'RE GOING TO BE HEROES.

The men were restless with anticipation. After all, they had been training in the deserts of New Mexico for months.

ENOLA GAY

Early the next morning, a film crew's lights illuminated the shining hull of the *Enola Gay,*

the B-29 bomber named after the mother of the pilot, Colonel Paul Tibbets.

As the crew performed their preflight checks, there was a sense among them that they were making history.

A little after midnight, the *Enola Gay* lumbered awkwardly into the skies and banked westward toward Japan.

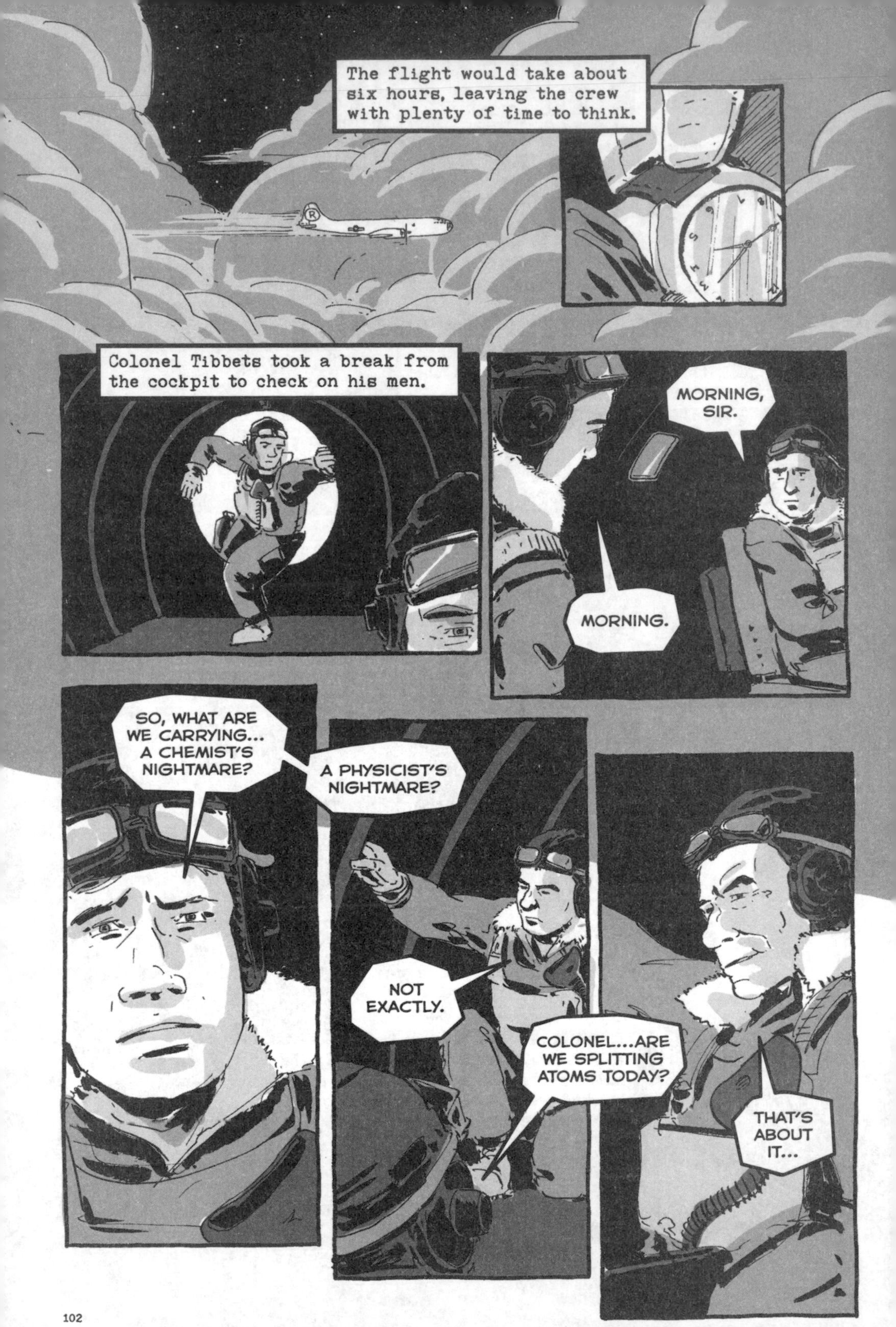
The flight would take about six hours, leaving the crew with plenty of time to think.
Colonel Tibbets took a break from the cockpit to check on his men.
MORNING, SIR.
MORNING.
SO, WHAT ARE WE CARRYING... A CHEMIST'S NIGHTMARE?
A PHYSICIST'S NIGHTMARE?
NOT EXACTLY.
COLONEL...ARE WE SPLITTING ATOMS TODAY?
THAT'S ABOUT IT...

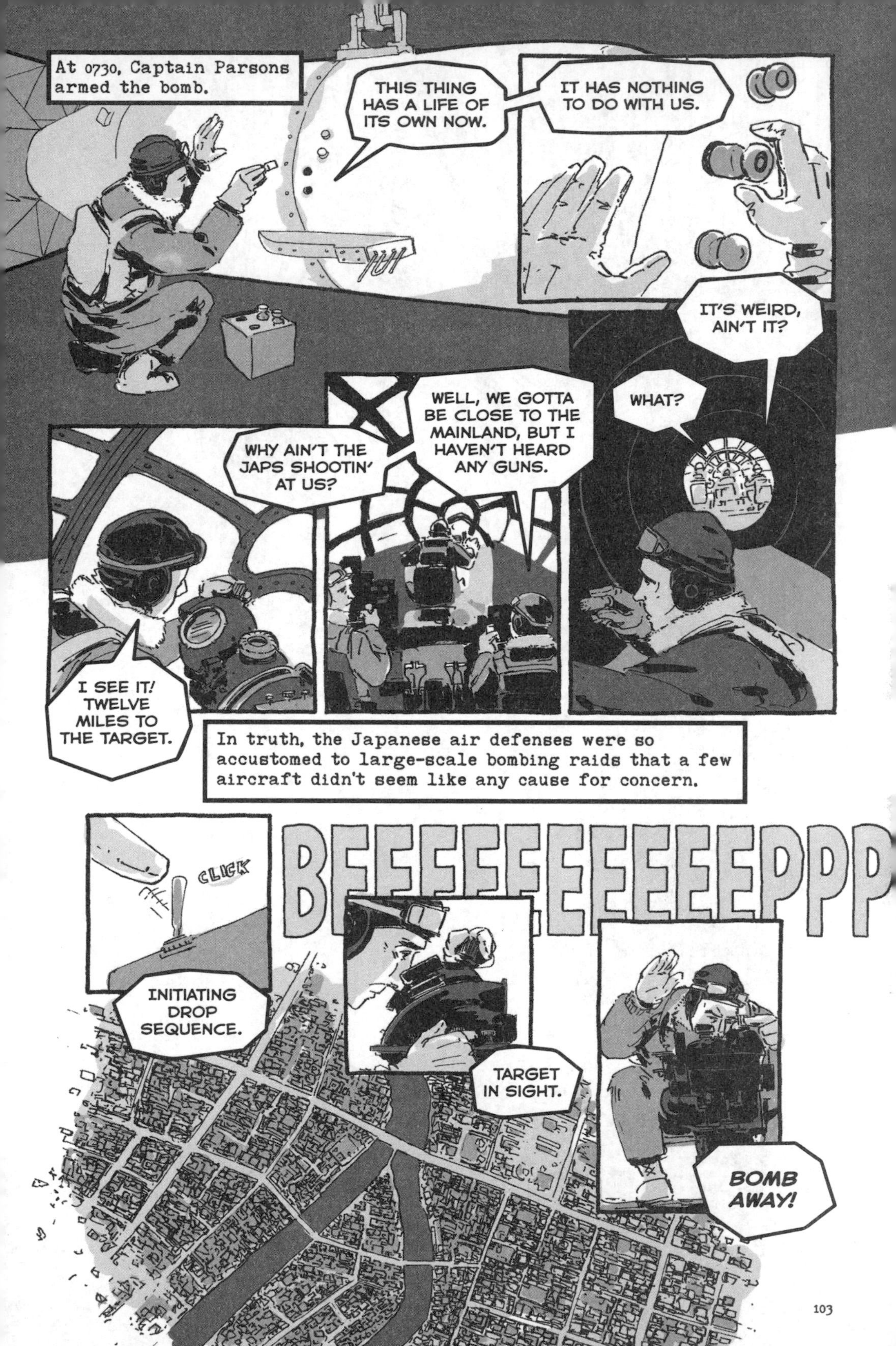
At 0730, Captain Parsons armed the bomb.
THIS THING HAS A LIFE OF ITS OWN NOW.
IT HAS NOTHING TO DO WITH US.
IT'S WEIRD, AIN'T IT?
WHAT?
WHY AIN'T THE JAPS SHOOTIN' AT US?
WELL, WE GOTTA BE CLOSE TO THE MAINLAND, BUT I HAVEN'T HEARD ANY GUNS.
I SEE IT! TWELVE MILES TO THE TARGET.
In truth, the Japanese air defenses were so accustomed to large-scale bombing raids that a few aircraft didn't seem like any cause for concern.
CLICK
BEEEEEEEEEEEPPP
INITIATING DROP SEQUENCE.
TARGET IN SIGHT.
BOMB AWAY!

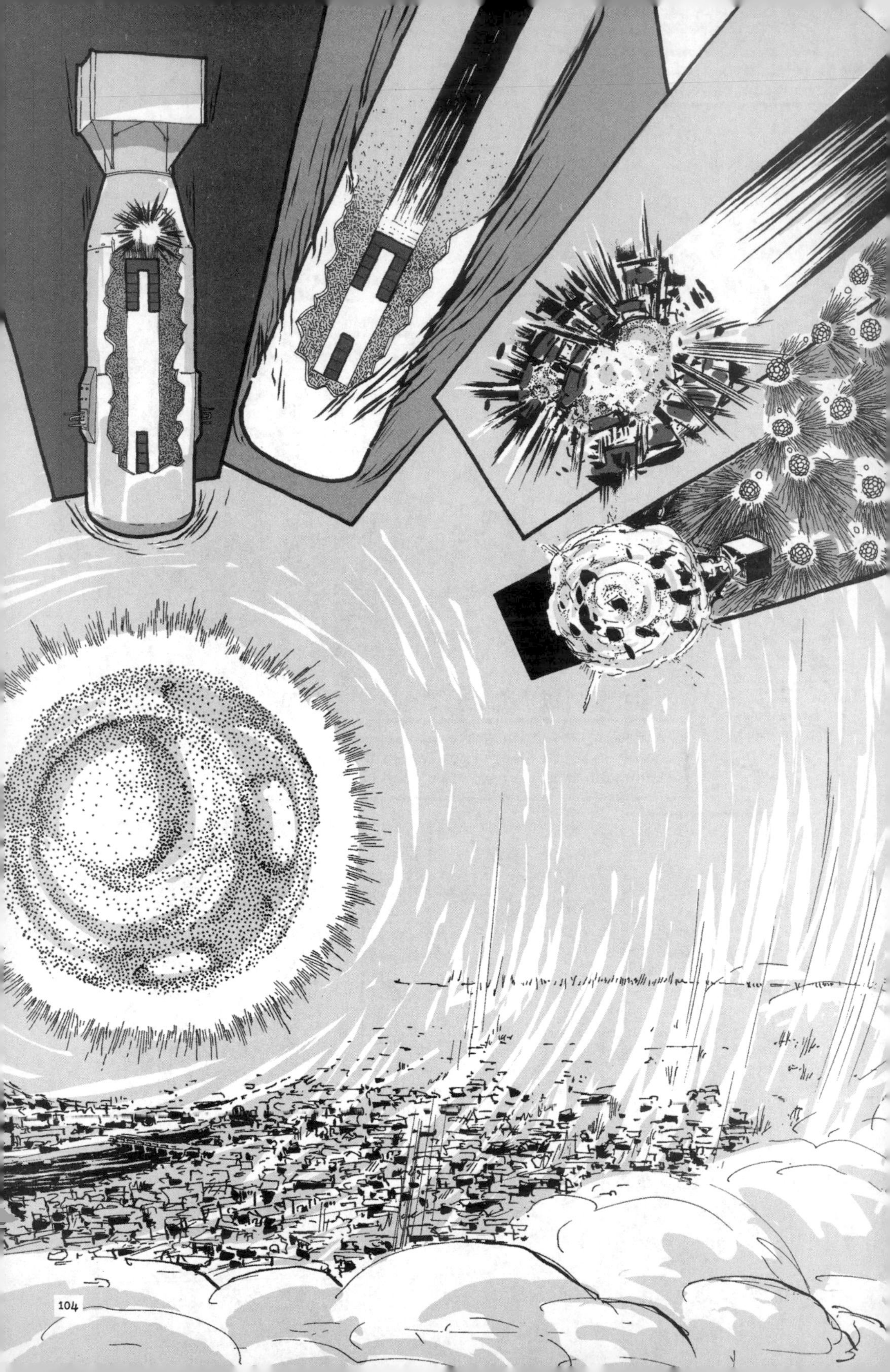

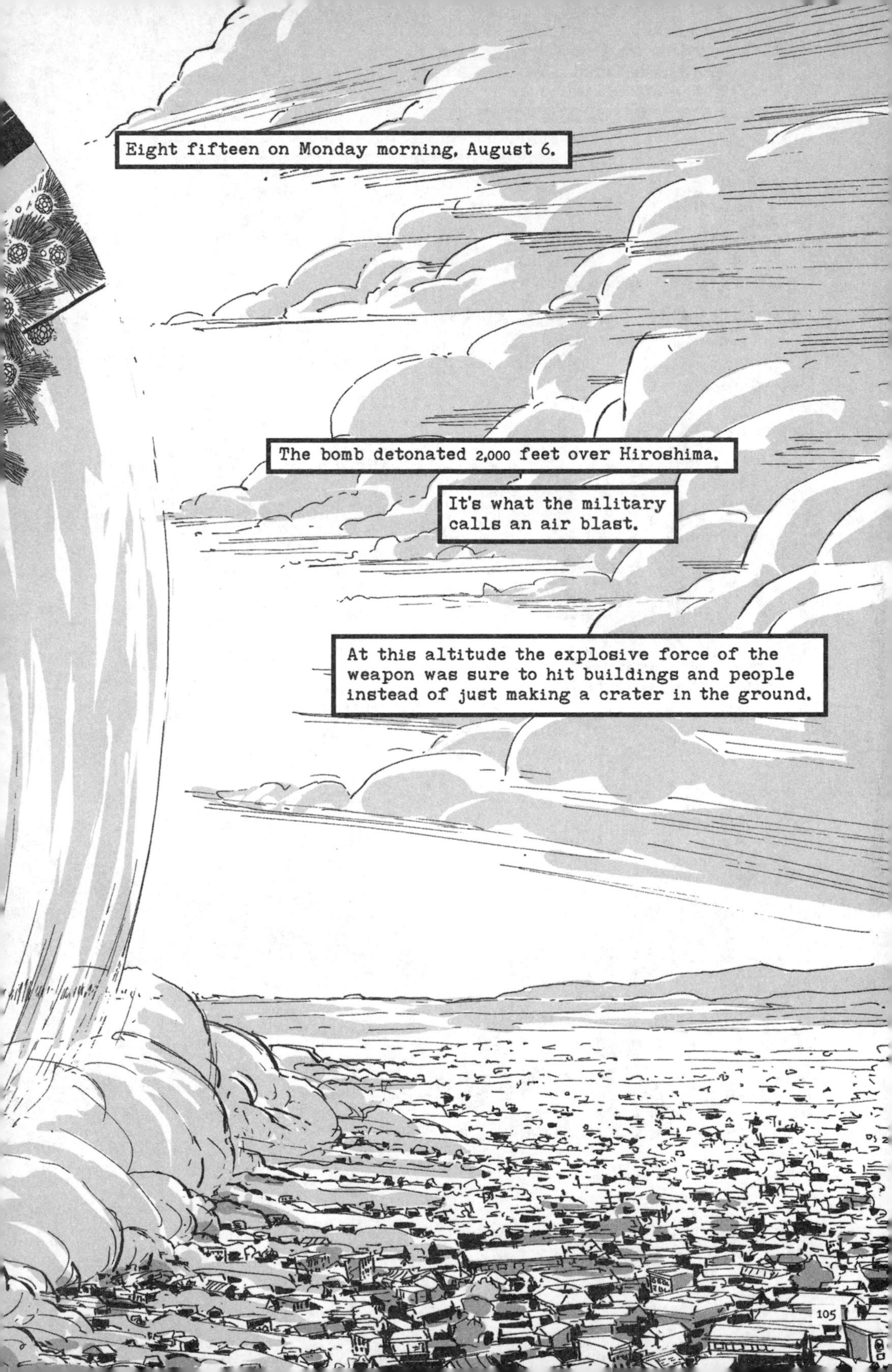
Eight fifteen on Monday morning, August 6.
The bomb detonated 2,000 feet over Hiroshima.
It's what the military calls an air blast.
At this altitude the explosive force of the weapon was sure to hit buildings and people instead of just making a crater in the ground.

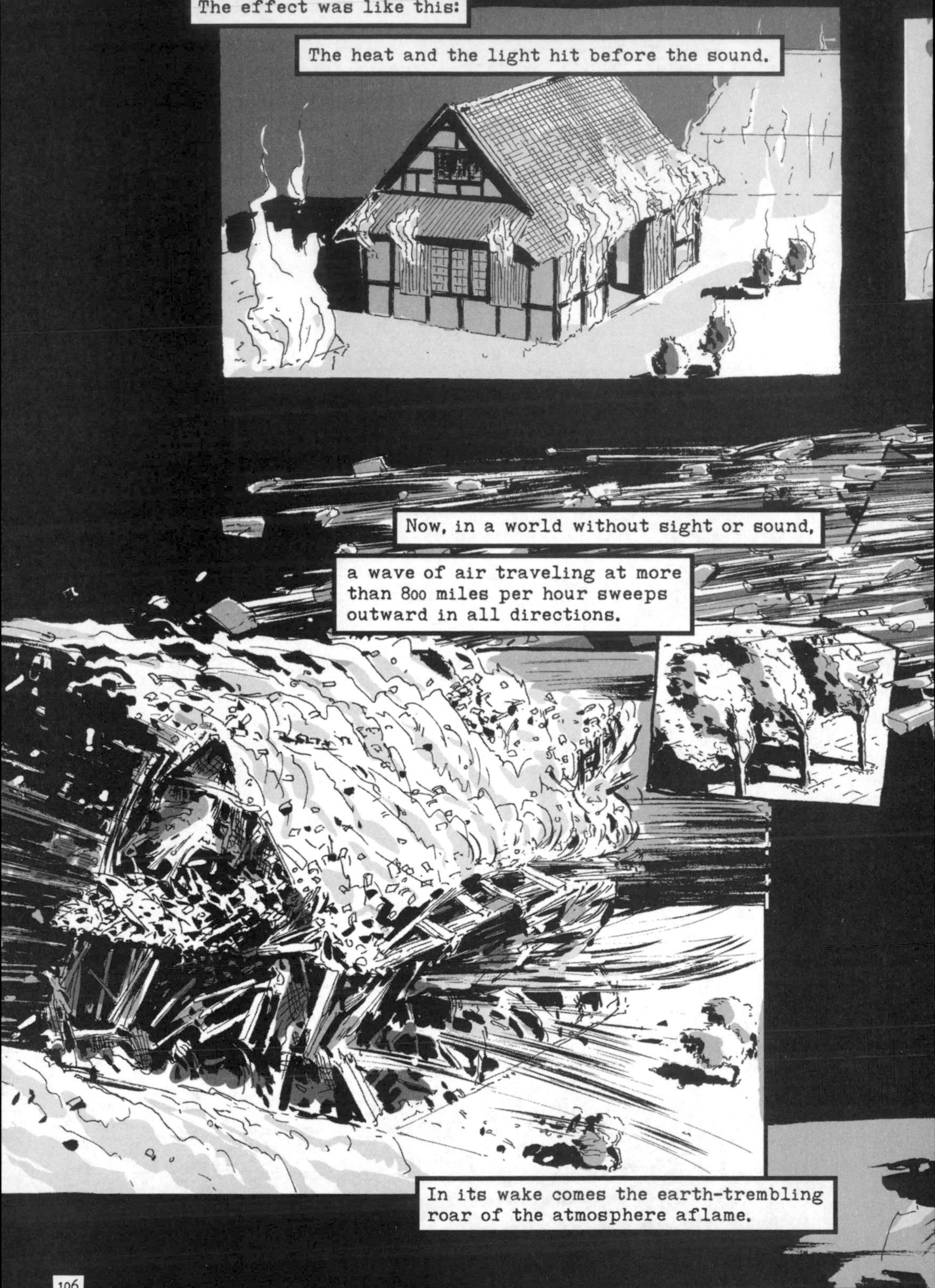
The effect was like this:
The heat and the light hit before the sound.
Now, in a world without sight or sound,
a wave of air traveling at more than 800 miles per hour sweeps outward in all directions.
In its wake comes the earth-trembling roar of the atmosphere aflame.

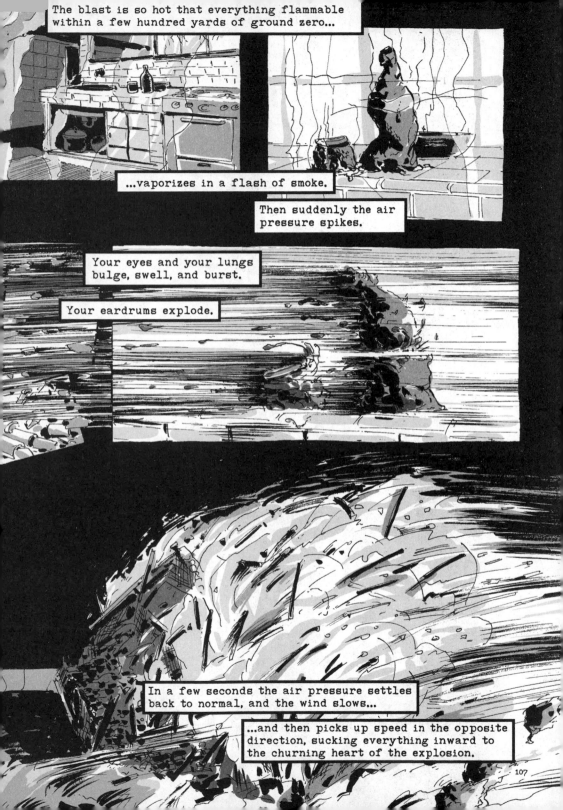

The blast is so hot that everything flammable within a few hundred yards of ground zero...

...vaporizes in a flash of smoke.

Then suddenly the air pressure spikes.

Your eyes and your lungs bulge, swell, and burst.

Your eardrums explode.

In a few seconds the air pressure settles back to normal, and the wind slows...

...and then picks up speed in the opposite direction, sucking everything inward to the churning heart of the explosion.

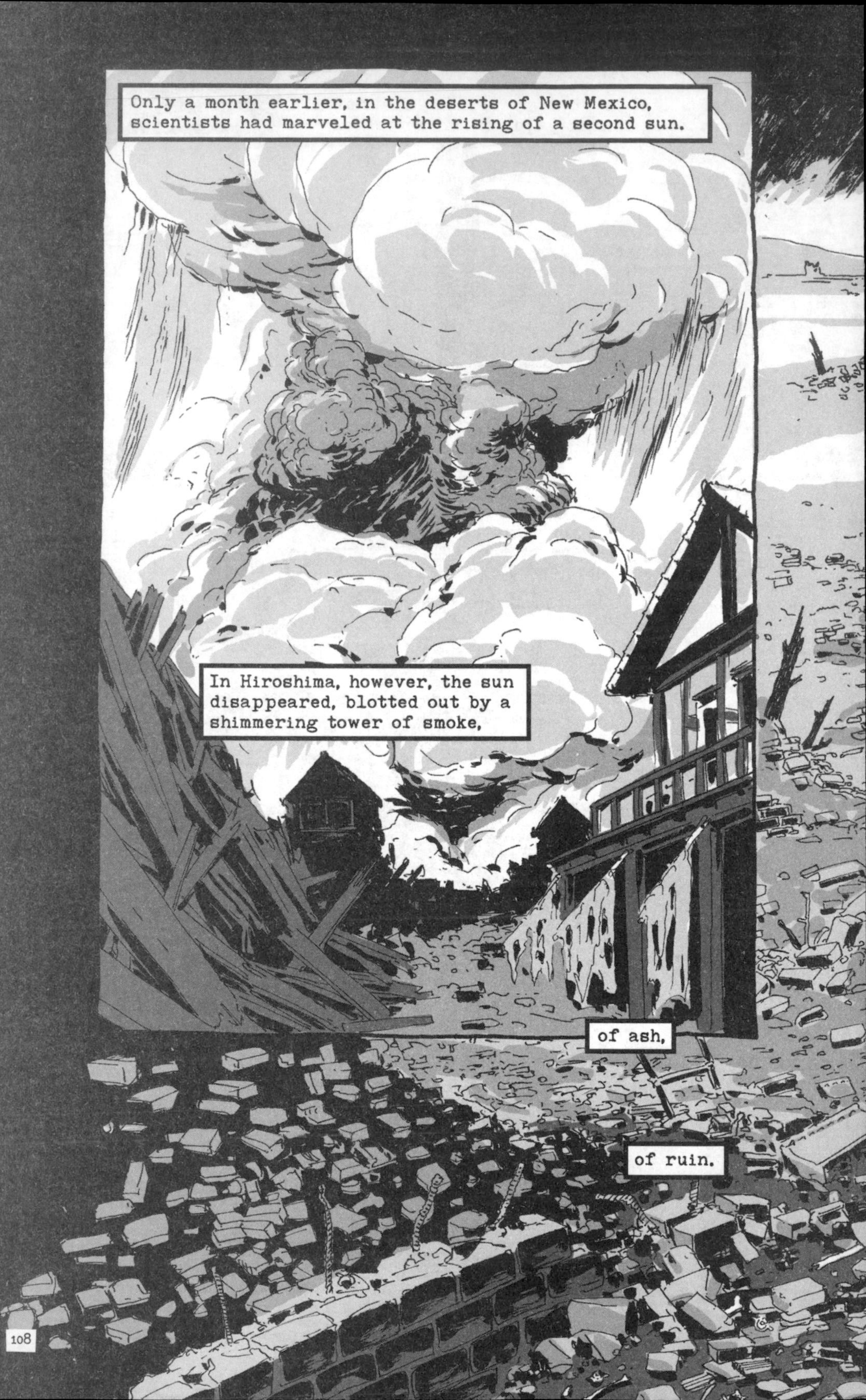
Only a month earlier, in the deserts of New Mexico, scientists had marveled at the rising of a second sun.
In Hiroshima, however, the sun disappeared, blotted out by a shimmering tower of smoke,
of ash,
of ruin.

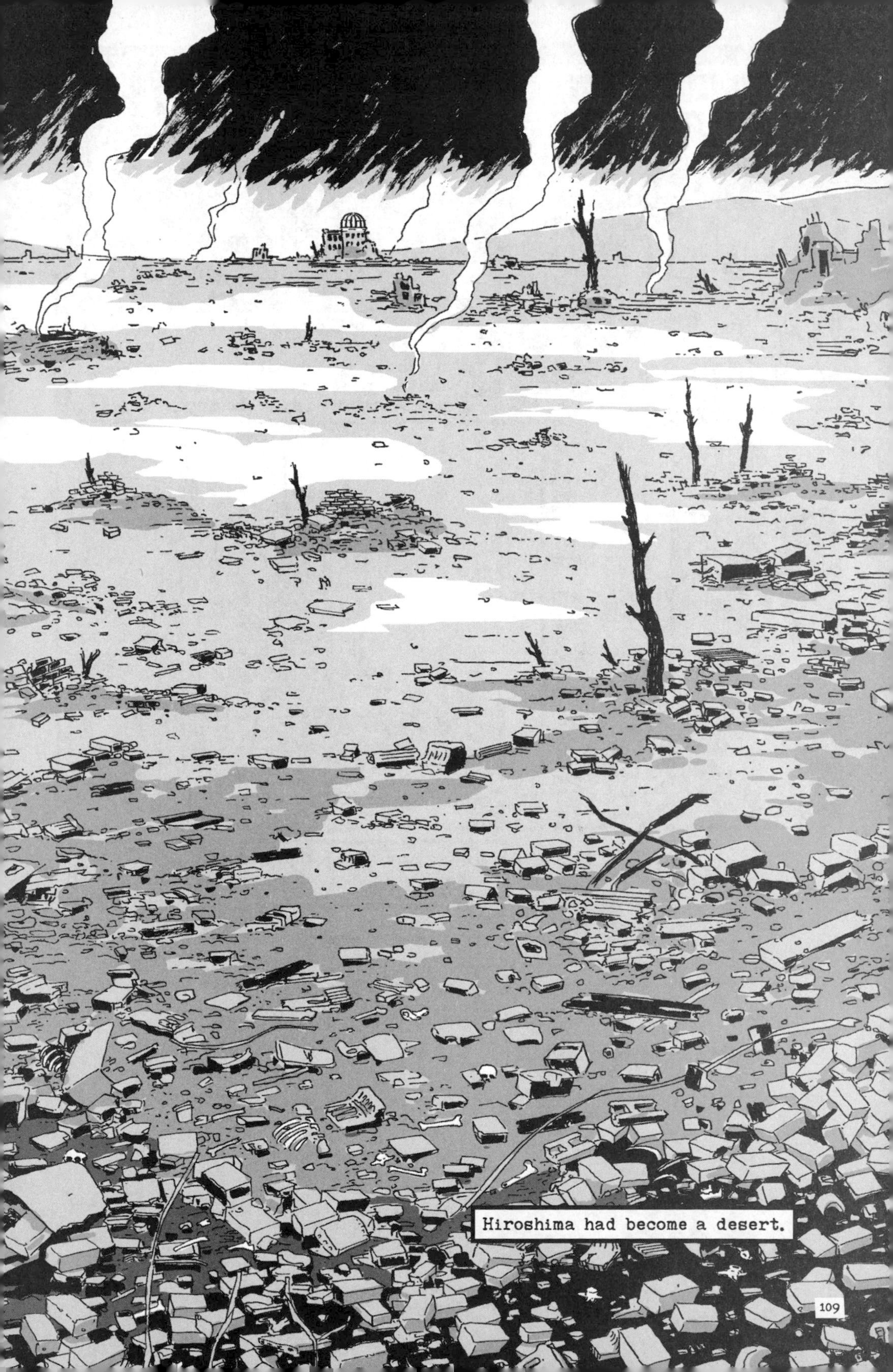
Hiroshima had become a desert.

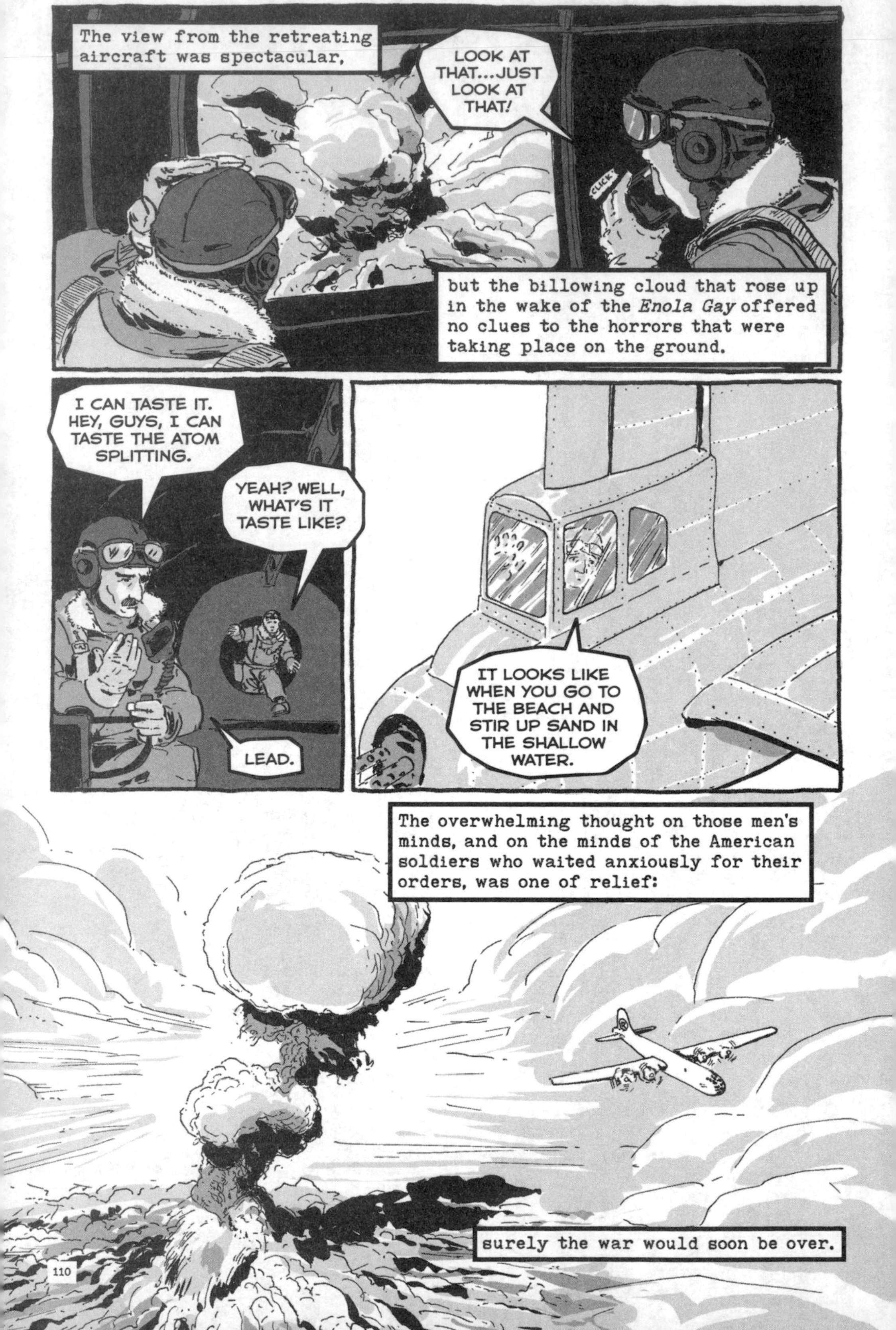
The view from the retreating aircraft was spectacular,
LOOK AT THAT...JUST LOOK AT THAT!
CLICK
but the billowing cloud that rose up in the wake of the *Enola Gay* offered no clues to the horrors that were taking place on the ground.
I CAN TASTE IT. HEY, GUYS, I CAN TASTE THE ATOM SPLITTING.
YEAH? WELL, WHAT'S IT TASTE LIKE?
LEAD.
IT LOOKS LIKE WHEN YOU GO TO THE BEACH AND STIR UP SAND IN THE SHALLOW WATER.
The overwhelming thought on those men's minds, and on the minds of the American soldiers who waited anxiously for their orders, was one of relief:
surely the war would soon be over.

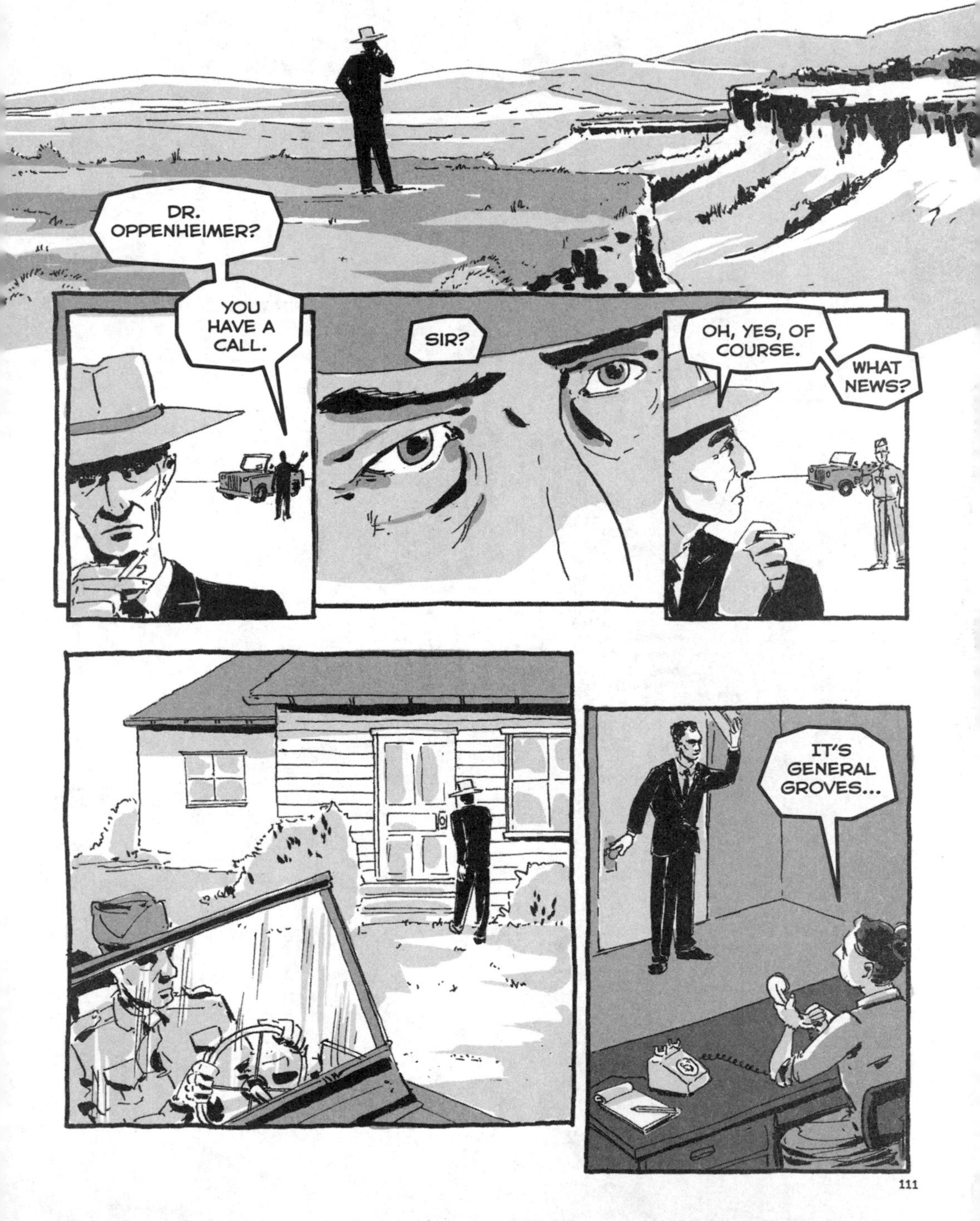
DR. OPPENHEIMER?
YOU HAVE A CALL.
SIR?
OH, YES, OF COURSE.
WHAT NEWS?
IT'S GENERAL GROVES...

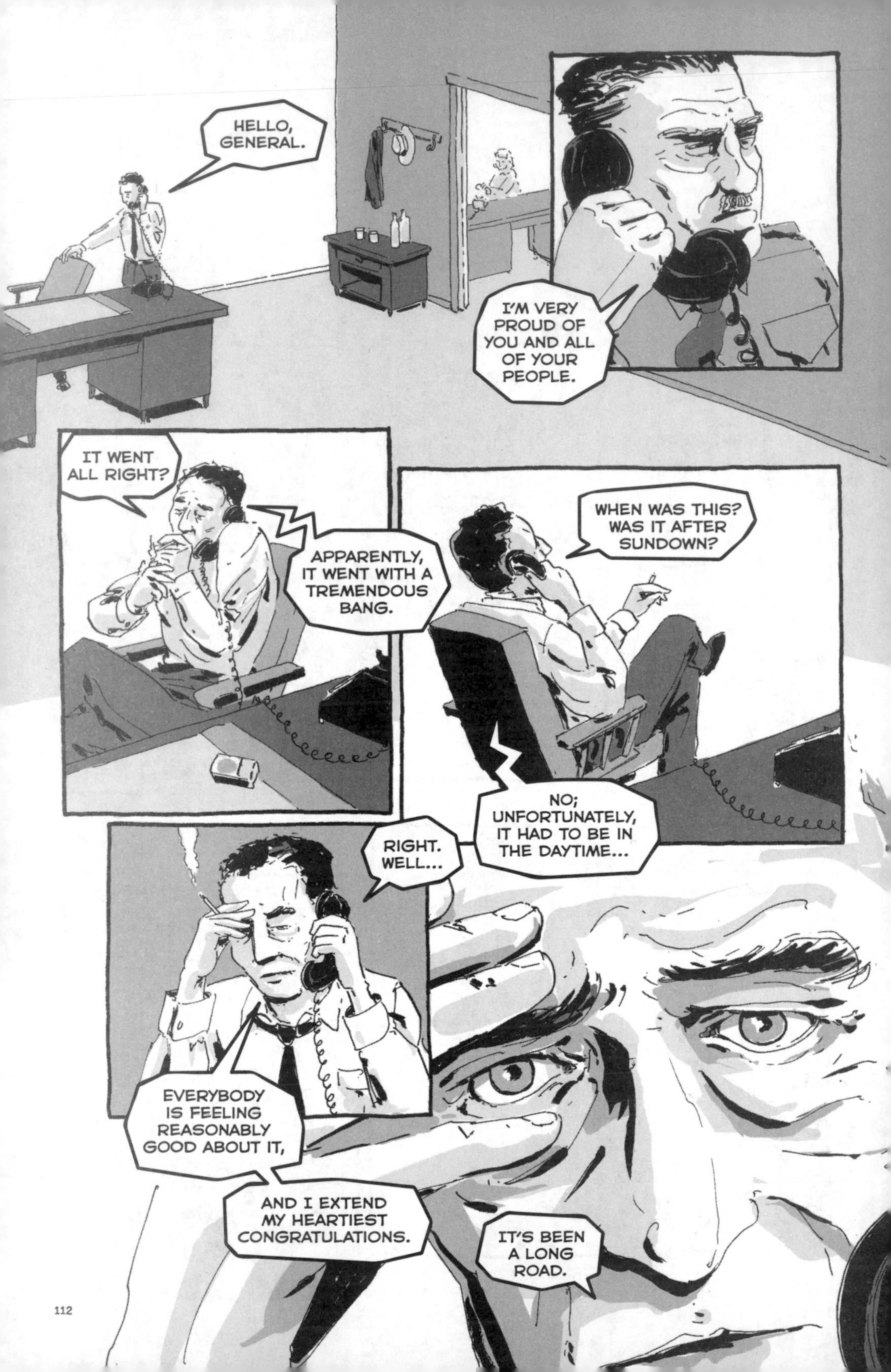
HELLO, GENERAL.
I'M VERY PROUD OF YOU AND ALL OF YOUR PEOPLE.
IT WENT ALL RIGHT?
APPARENTLY, IT WENT WITH A TREMENDOUS BANG.
WHEN WAS THIS? WAS IT AFTER SUNDOWN?
NO; UNFORTUNATELY, IT HAD TO BE IN THE DAYTIME...
RIGHT. WELL...
EVERYBODY IS FEELING REASONABLY GOOD ABOUT IT,
AND I EXTEND MY HEARTIEST CONGRATULATIONS.
IT'S BEEN A LONG ROAD.

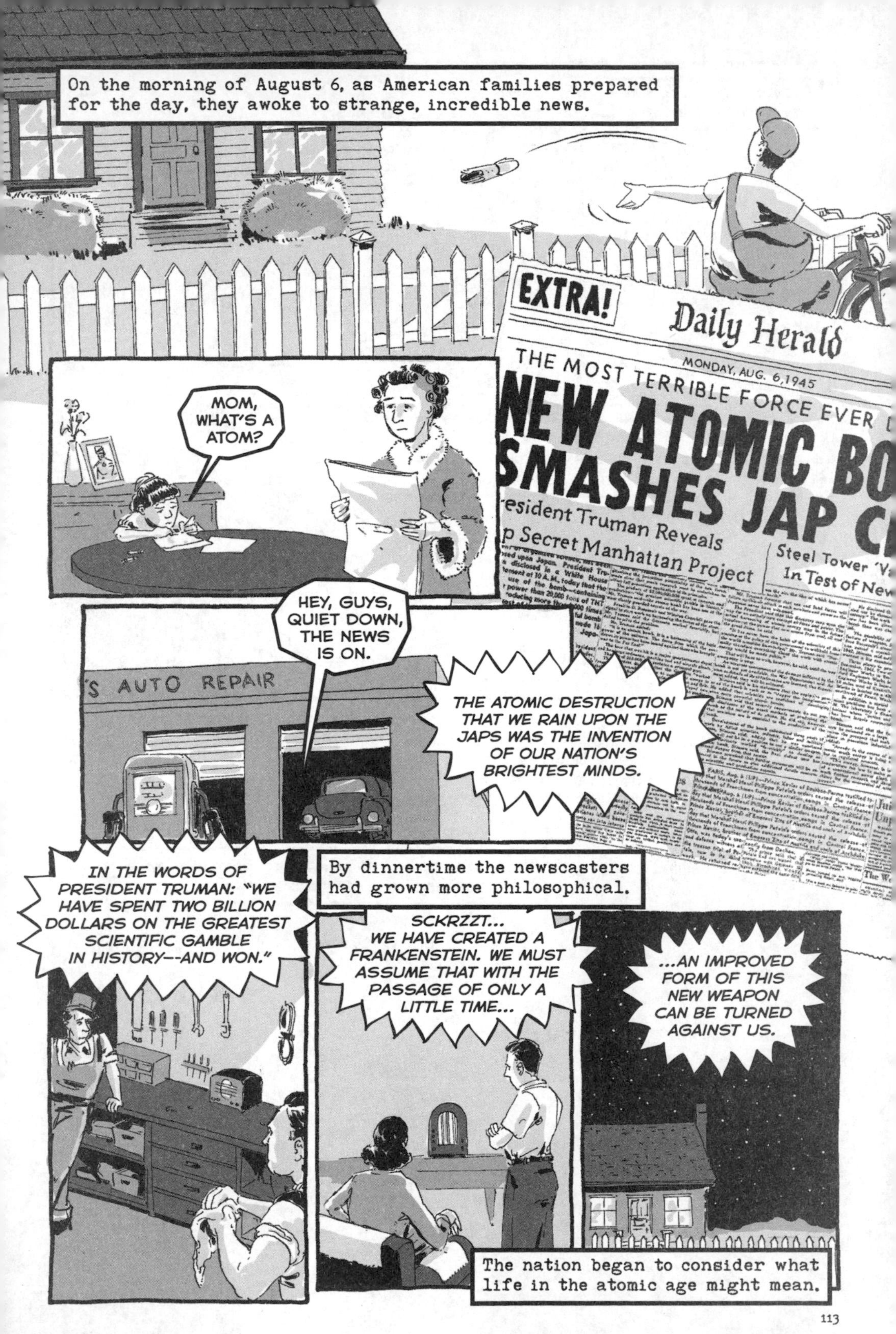

On the morning of August 6, as American families prepared for the day, they awoke to strange, incredible news.
EXTRA!
Daily Herald
MONDAY, AUG. 6, 1945
THE MOST TERRIBLE FORCE EVER
NEW ATOMIC BO
SMASHES JAP C
resident Truman Reveals
p Secret Manhattan Project
Steel Tower 'V
In Test of New
MOM, WHAT'S A ATOM?
HEY, GUYS, QUIET DOWN, THE NEWS IS ON.
'S AUTO REPAIR
THE ATOMIC DESTRUCTION THAT WE RAIN UPON THE JAPS WAS THE INVENTION OF OUR NATION'S BRIGHTEST MINDS.
IN THE WORDS OF PRESIDENT TRUMAN: "WE HAVE SPENT TWO BILLION DOLLARS ON THE GREATEST SCIENTIFIC GAMBLE IN HISTORY--AND WON."
By dinnertime the newscasters had grown more philosophical.
SCKRZZT... WE HAVE CREATED A FRANKENSTEIN. WE MUST ASSUME THAT WITH THE PASSAGE OF ONLY A LITTLE TIME...
...AN IMPROVED FORM OF THIS NEW WEAPON CAN BE TURNED AGAINST US.
The nation began to consider what life in the atomic age might mean.

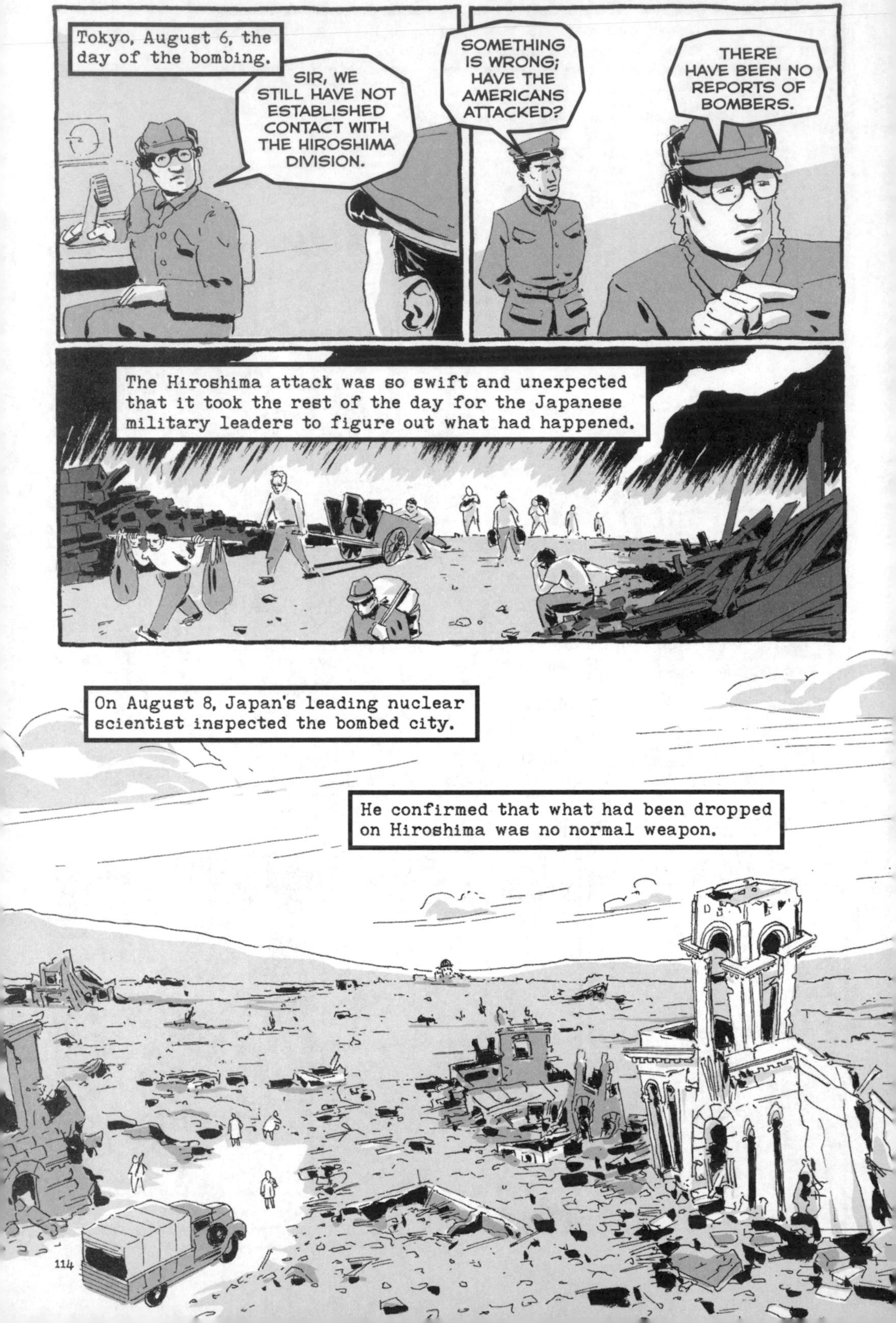
Tokyo, August 6, the day of the bombing.
SIR, WE STILL HAVE NOT ESTABLISHED CONTACT WITH THE HIROSHIMA DIVISION.
SOMETHING IS WRONG; HAVE THE AMERICANS ATTACKED?
THERE HAVE BEEN NO REPORTS OF BOMBERS.
The Hiroshima attack was so swift and unexpected that it took the rest of the day for the Japanese military leaders to figure out what had happened.
On August 8, Japan's leading nuclear scientist inspected the bombed city.
He confirmed that what had been dropped on Hiroshima was no normal weapon.

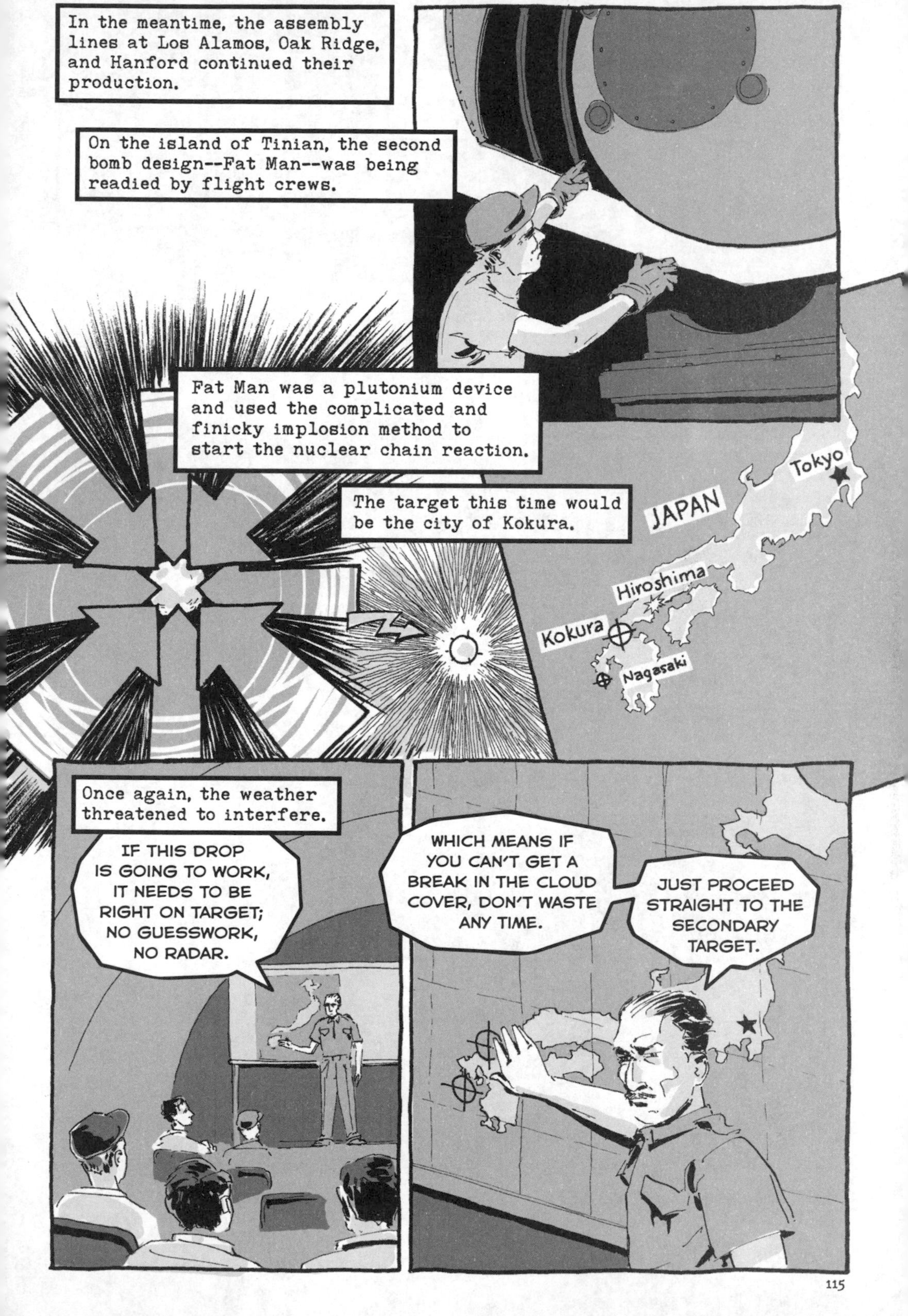
In the meantime, the assembly lines at Los Alamos, Oak Ridge, and Hanford continued their production.
On the island of Tinian, the second bomb design--Fat Man--was being readied by flight crews.
Fat Man was a plutonium device and used the complicated and finicky implosion method to start the nuclear chain reaction.
The target this time would be the city of Kokura.
JAPAN
Tokyo
Hiroshima
Kokura
Nagasaki
Once again, the weather threatened to interfere.
IF THIS DROP IS GOING TO WORK, IT NEEDS TO BE RIGHT ON TARGET; NO GUESSWORK, NO RADAR.
WHICH MEANS IF YOU CAN'T GET A BREAK IN THE CLOUD COVER, DON'T WASTE ANY TIME.
JUST PROCEED STRAIGHT TO THE SECONDARY TARGET.

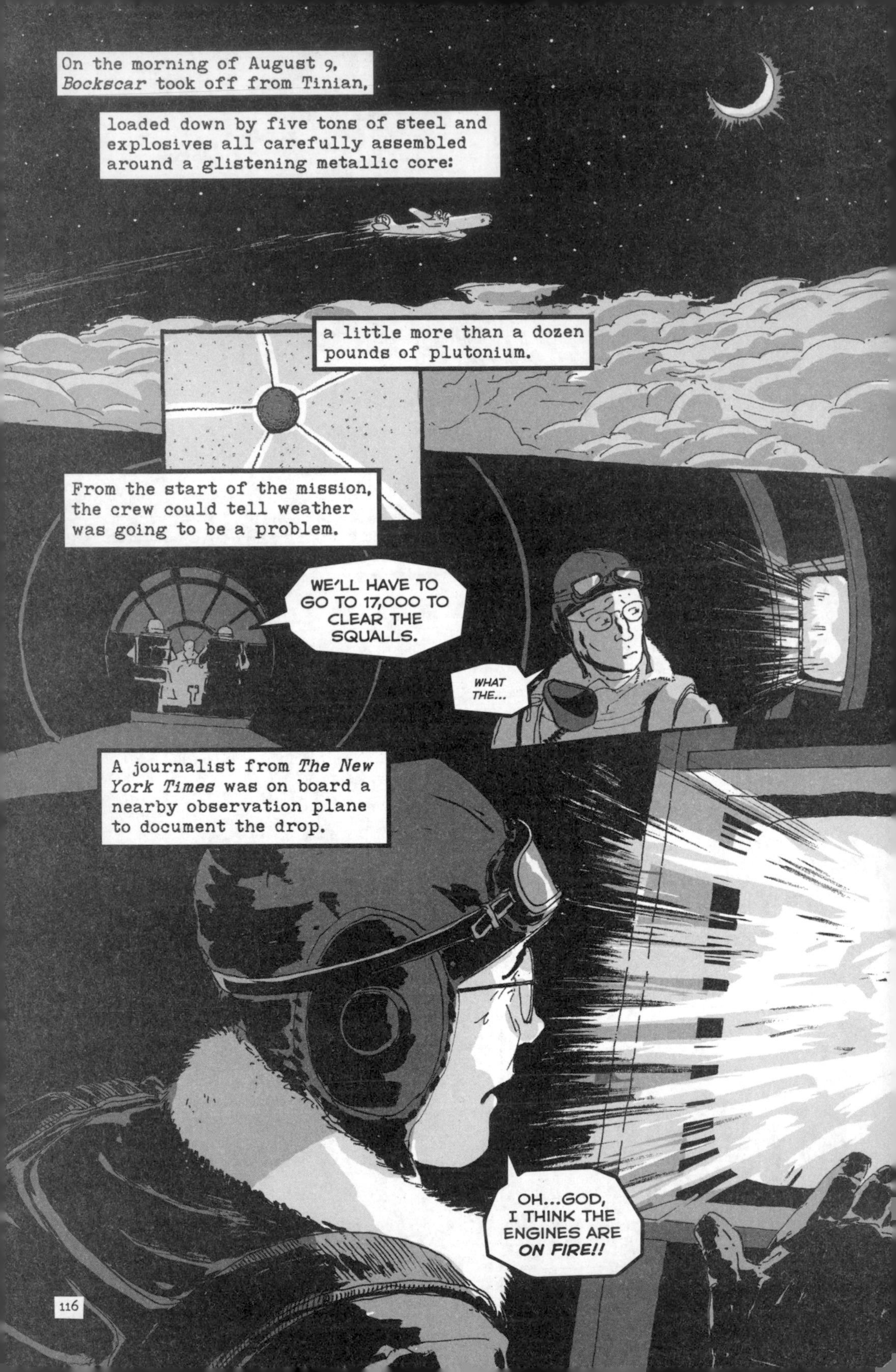
On the morning of August 9, *Bockscar* took off from Tinian,
loaded down by five tons of steel and explosives all carefully assembled around a glistening metallic core:
a little more than a dozen pounds of plutonium.
From the start of the mission, the crew could tell weather was going to be a problem.
WE'LL HAVE TO GO TO 17,000 TO CLEAR THE SQUALLS.
WHAT THE...
A journalist from *The New York Times* was on board a nearby observation plane to document the drop.
OH...GOD, I THINK THE ENGINES ARE ***ON FIRE!!***

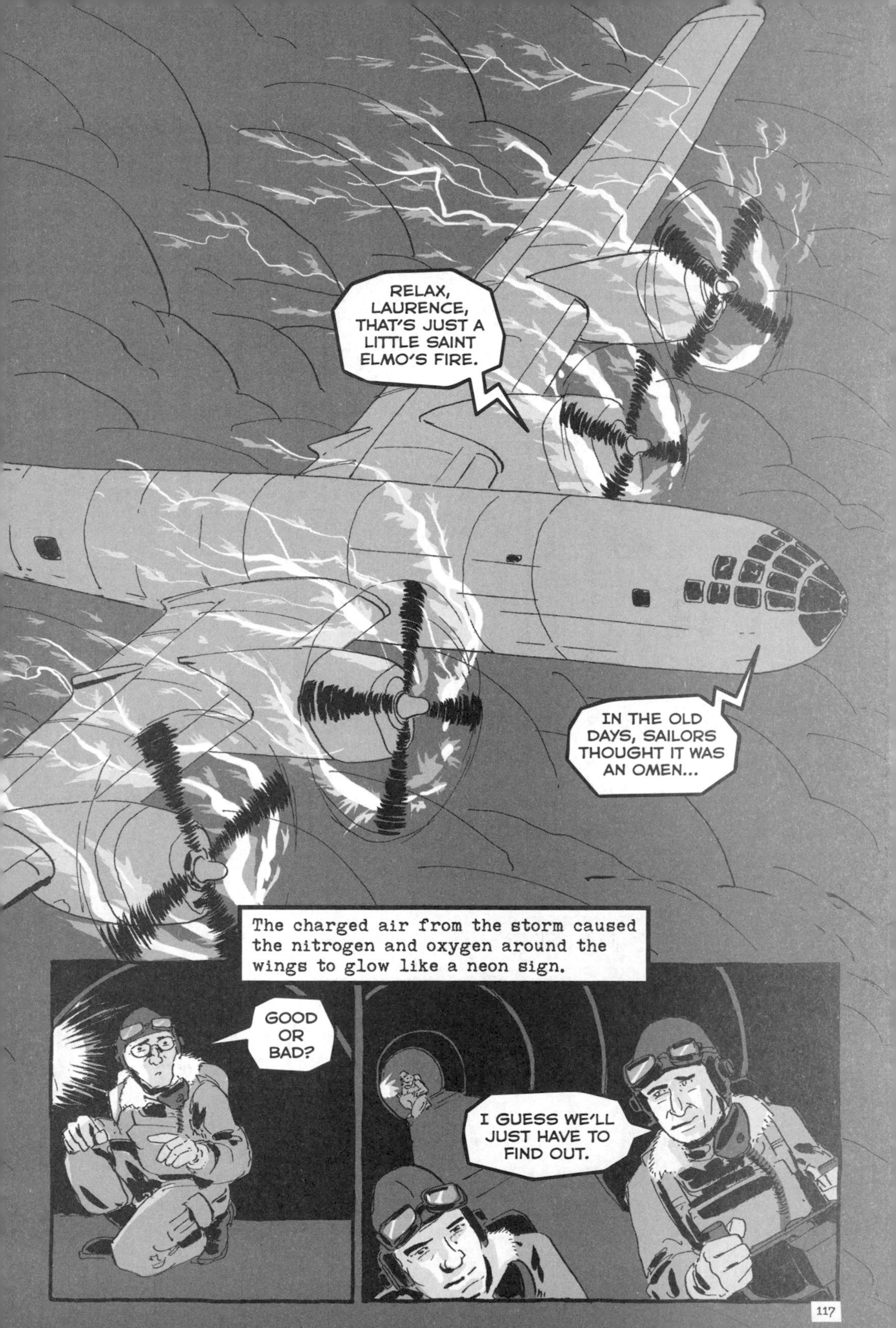
RELAX, LAURENCE, THAT'S JUST A LITTLE SAINT ELMO'S FIRE.
IN THE OLD DAYS, SAILORS THOUGHT IT WAS AN OMEN...
The charged air from the storm caused the nitrogen and oxygen around the wings to glow like a neon sign.
GOOD OR BAD?
I GUESS WE'LL JUST HAVE TO FIND OUT.

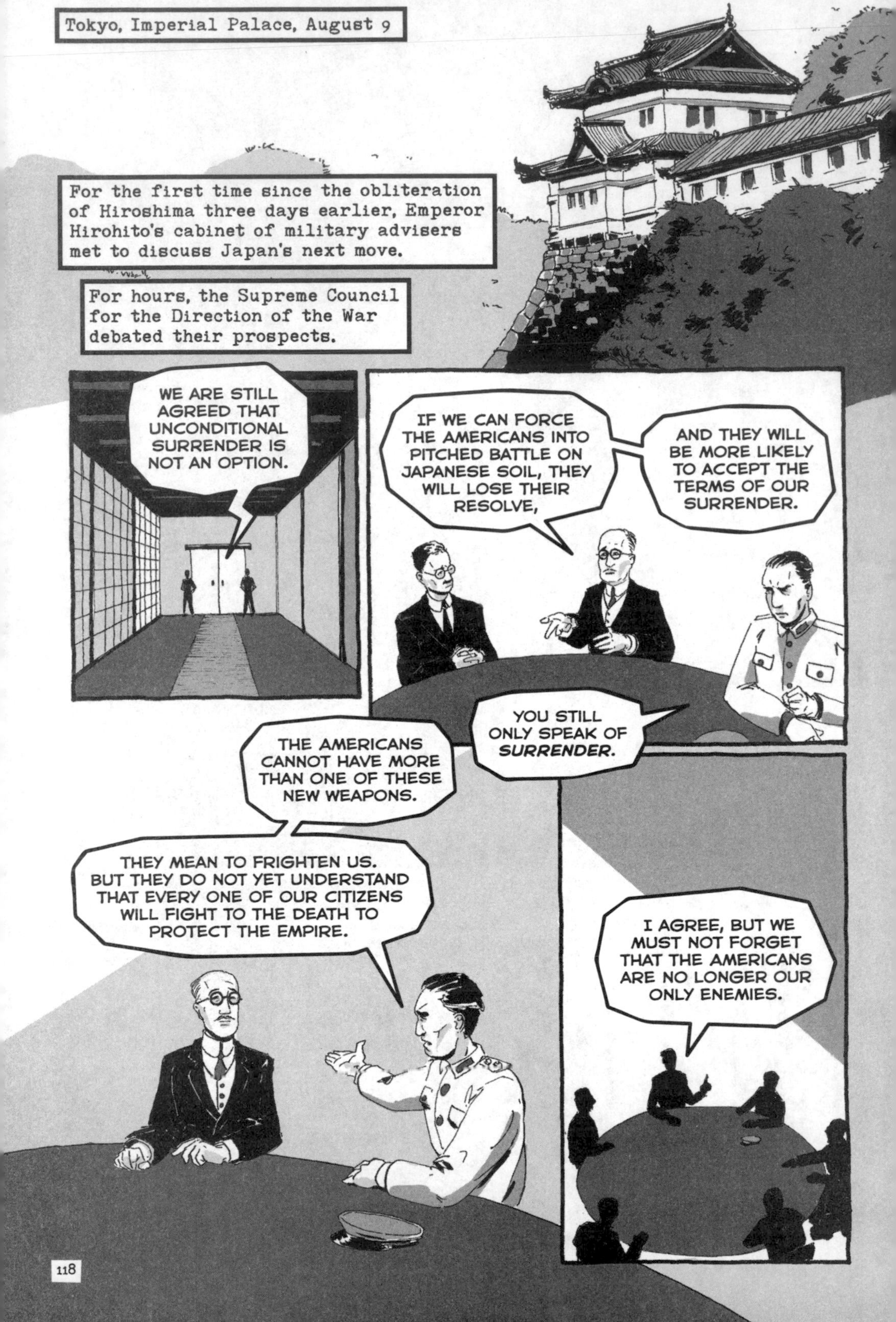

Tokyo, Imperial Palace, August 9
For the first time since the obliteration of Hiroshima three days earlier, Emperor Hirohito's cabinet of military advisers met to discuss Japan's next move.
For hours, the Supreme Council for the Direction of the War debated their prospects.
WE ARE STILL AGREED THAT UNCONDITIONAL SURRENDER IS NOT AN OPTION.
IF WE CAN FORCE THE AMERICANS INTO PITCHED BATTLE ON JAPANESE SOIL, THEY WILL LOSE THEIR RESOLVE,
AND THEY WILL BE MORE LIKELY TO ACCEPT THE TERMS OF OUR SURRENDER.
YOU STILL ONLY SPEAK OF *SURRENDER*.
THE AMERICANS CANNOT HAVE MORE THAN ONE OF THESE NEW WEAPONS.
THEY MEAN TO FRIGHTEN US. BUT THEY DO NOT YET UNDERSTAND THAT EVERY ONE OF OUR CITIZENS WILL FIGHT TO THE DEATH TO PROTECT THE EMPIRE.
I AGREE, BUT WE MUST NOT FORGET THAT THE AMERICANS ARE NO LONGER OUR ONLY ENEMIES.

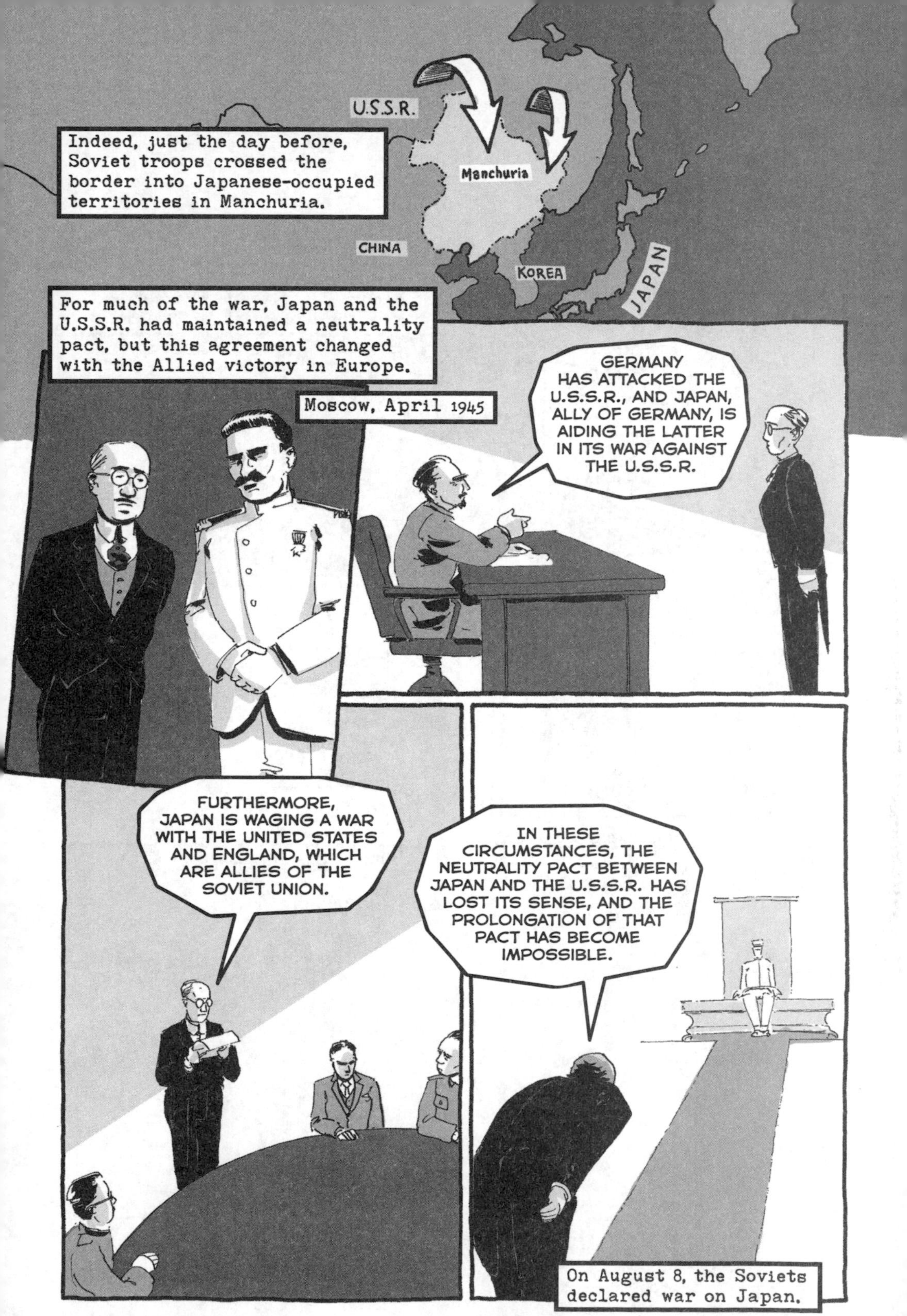

U.S.S.R.
Indeed, just the day before, Soviet troops crossed the border into Japanese-occupied territories in Manchuria.
Manchuria
CHINA
KOREA
JAPAN
For much of the war, Japan and the U.S.S.R. had maintained a neutrality pact, but this agreement changed with the Allied victory in Europe.
Moscow, April 1945
GERMANY HAS ATTACKED THE U.S.S.R., AND JAPAN, ALLY OF GERMANY, IS AIDING THE LATTER IN ITS WAR AGAINST THE U.S.S.R.
FURTHERMORE, JAPAN IS WAGING A WAR WITH THE UNITED STATES AND ENGLAND, WHICH ARE ALLIES OF THE SOVIET UNION.
IN THESE CIRCUMSTANCES, THE NEUTRALITY PACT BETWEEN JAPAN AND THE U.S.S.R. HAS LOST ITS SENSE, AND THE PROLONGATION OF THAT PACT HAS BECOME IMPOSSIBLE.
On August 8, the Soviets declared war on Japan.

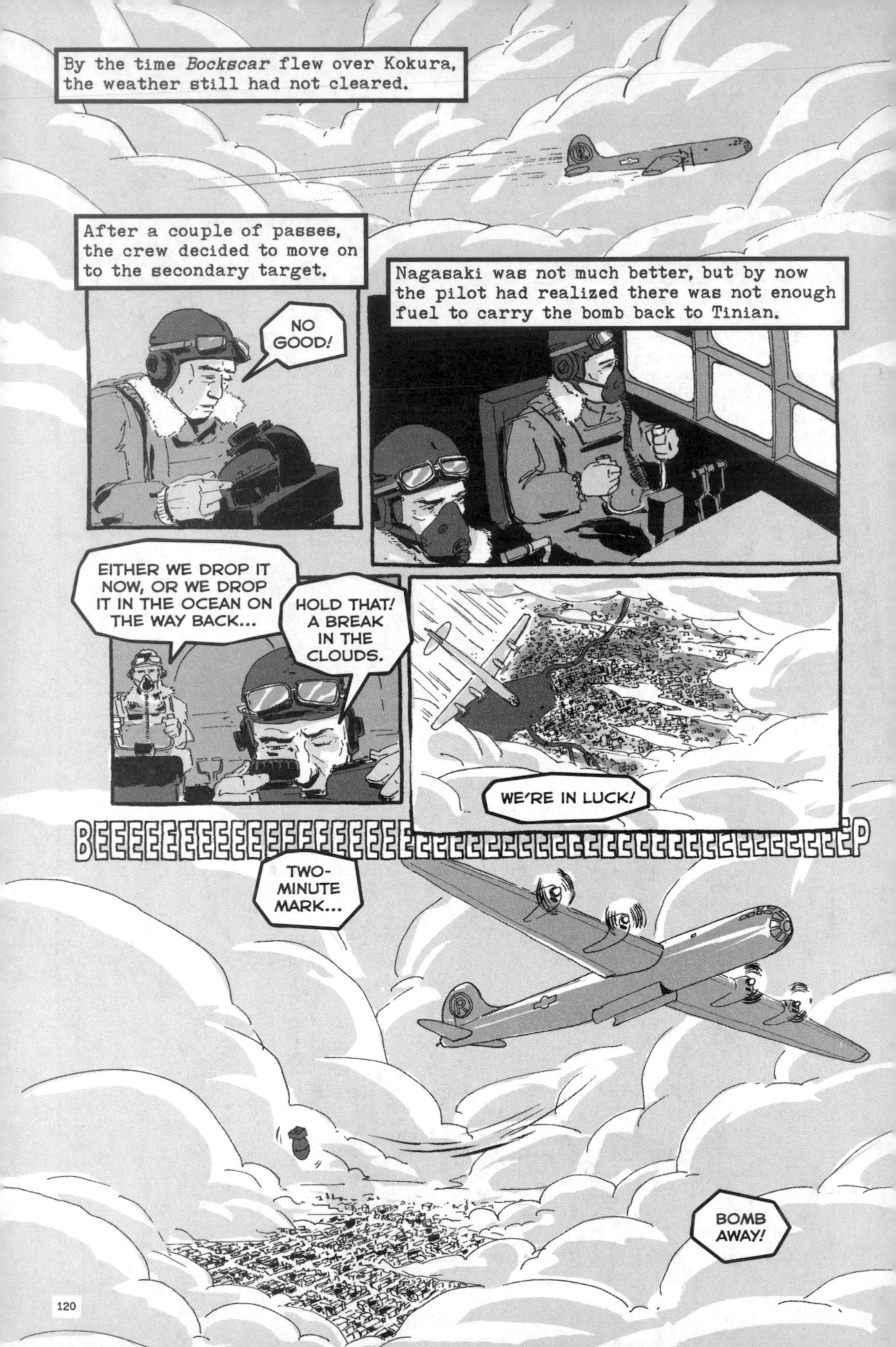

By the time *Bockscar* flew over Kokura, the weather still had not cleared.
After a couple of passes, the crew decided to move on to the secondary target.
NO GOOD!
Nagasaki was not much better, but by now the pilot had realized there was not enough fuel to carry the bomb back to Tinian.
EITHER WE DROP IT NOW, OR WE DROP IT IN THE OCEAN ON THE WAY BACK...
HOLD THAT! A BREAK IN THE CLOUDS.
WE'RE IN LUCK!
BEEEEEEEEEEEEEEEEEEEEEEEEEEEEEEEEEEEEEEEEEEEEEP
TWO-MINUTE MARK...
BOMB AWAY!

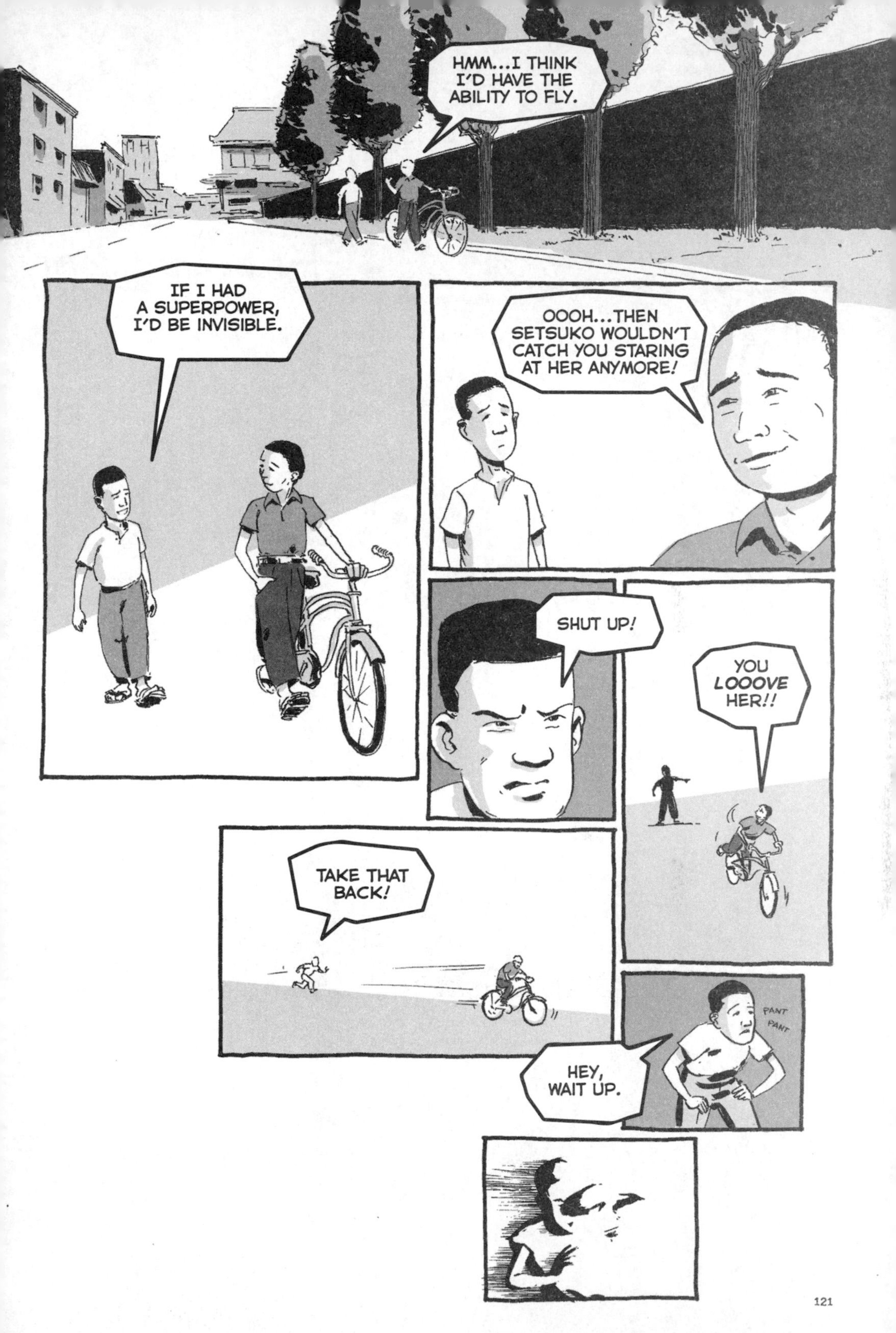
HMM...I THINK I'D HAVE THE ABILITY TO FLY.
IF I HAD A SUPERPOWER, I'D BE INVISIBLE.
OOOH...THEN SETSUKO WOULDN'T CATCH YOU STARING AT HER ANYMORE!
SHUT UP!
YOU LOOOVE HER!!
TAKE THAT BACK!
PANT PANT
HEY, WAIT UP.

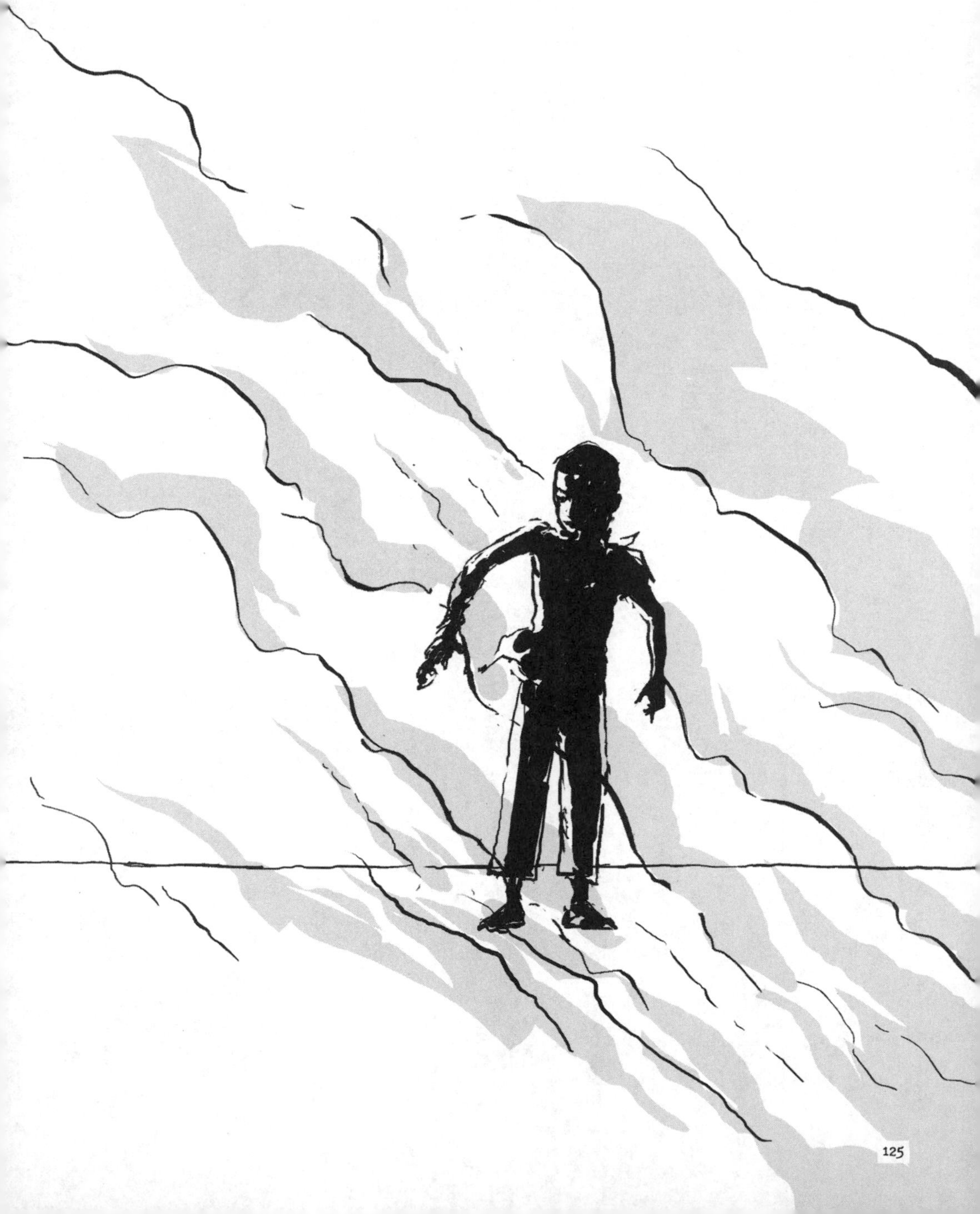

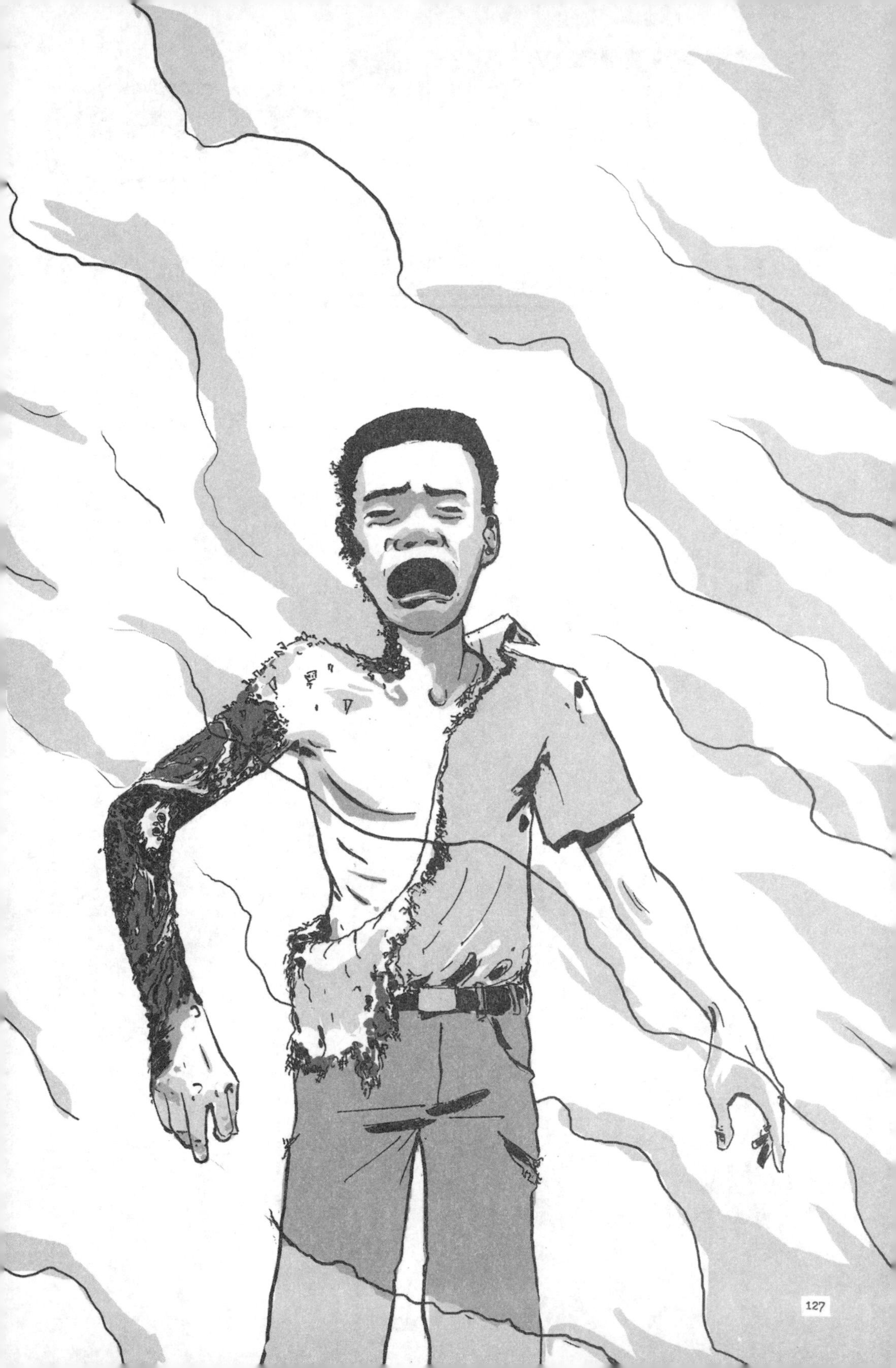

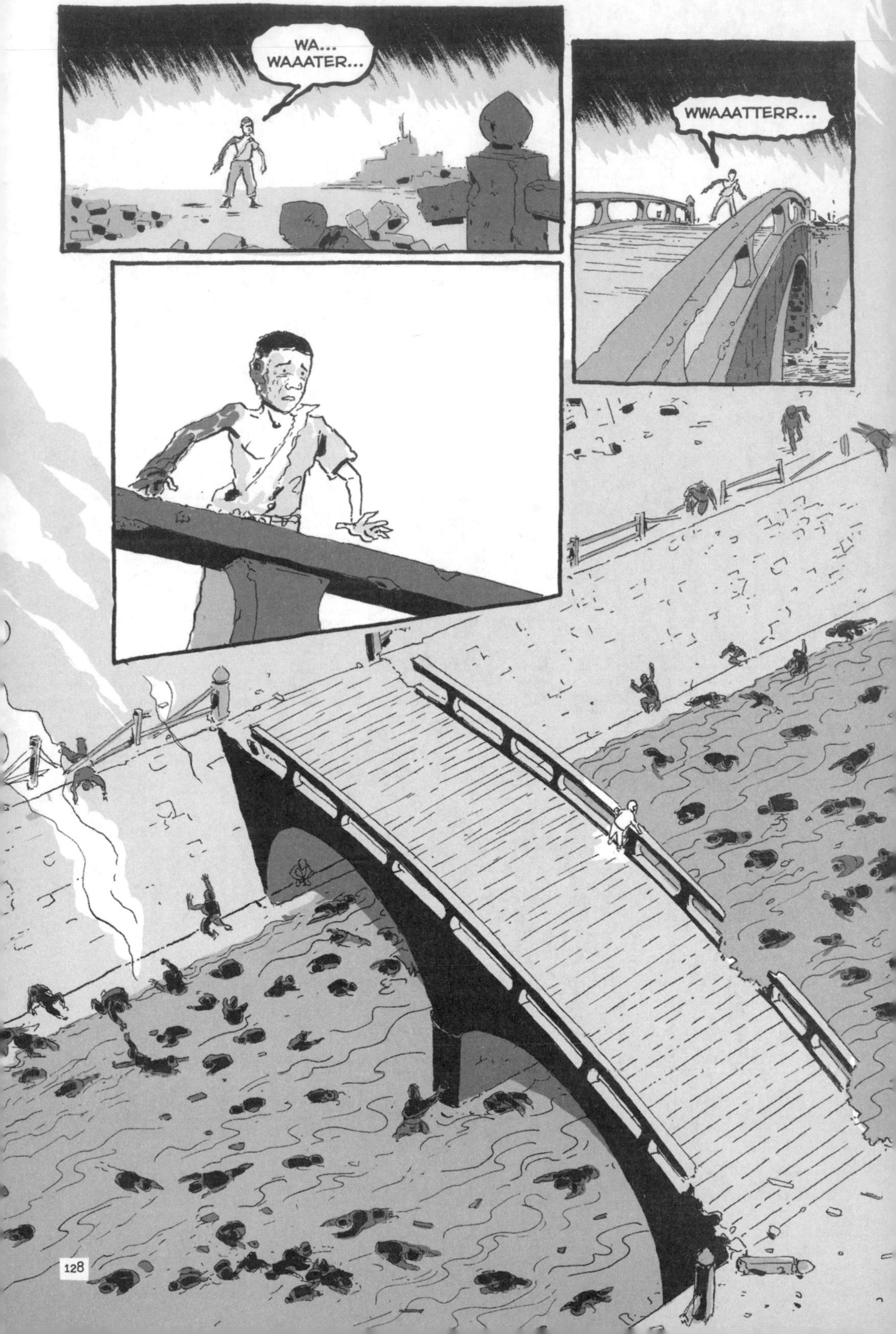
WA...
WAAATER...
WWAAATTERR...

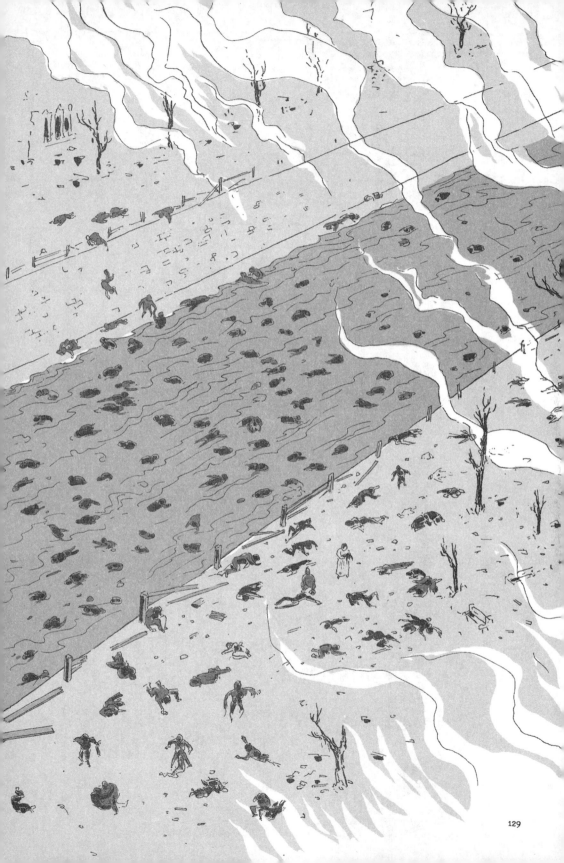

129

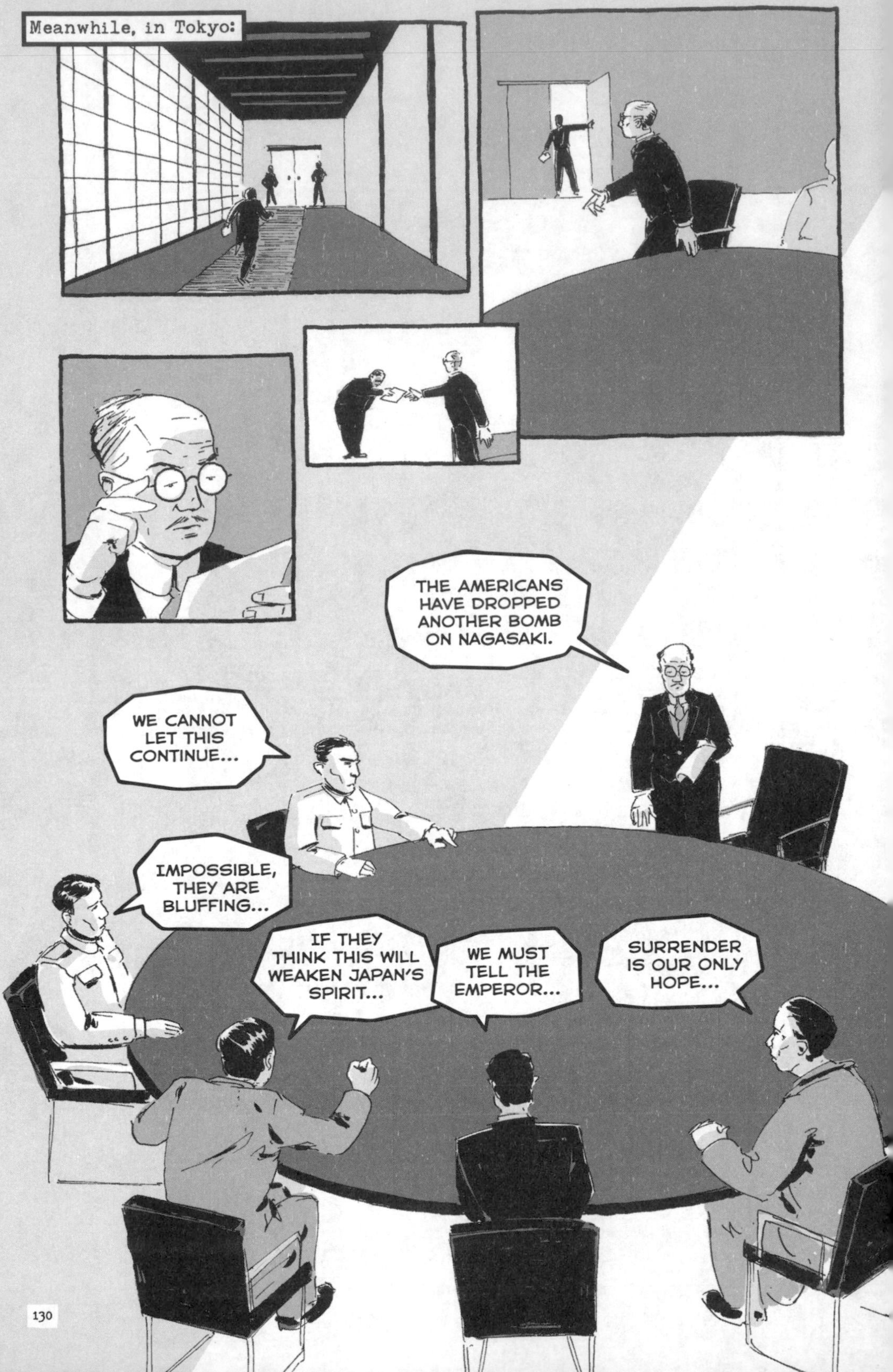
Meanwhile, in Tokyo:
THE AMERICANS HAVE DROPPED ANOTHER BOMB ON NAGASAKI.
WE CANNOT LET THIS CONTINUE...
IMPOSSIBLE, THEY ARE BLUFFING...
IF THEY THINK THIS WILL WEAKEN JAPAN'S SPIRIT...
WE MUST TELL THE EMPEROR...
SURRENDER IS OUR ONLY HOPE...

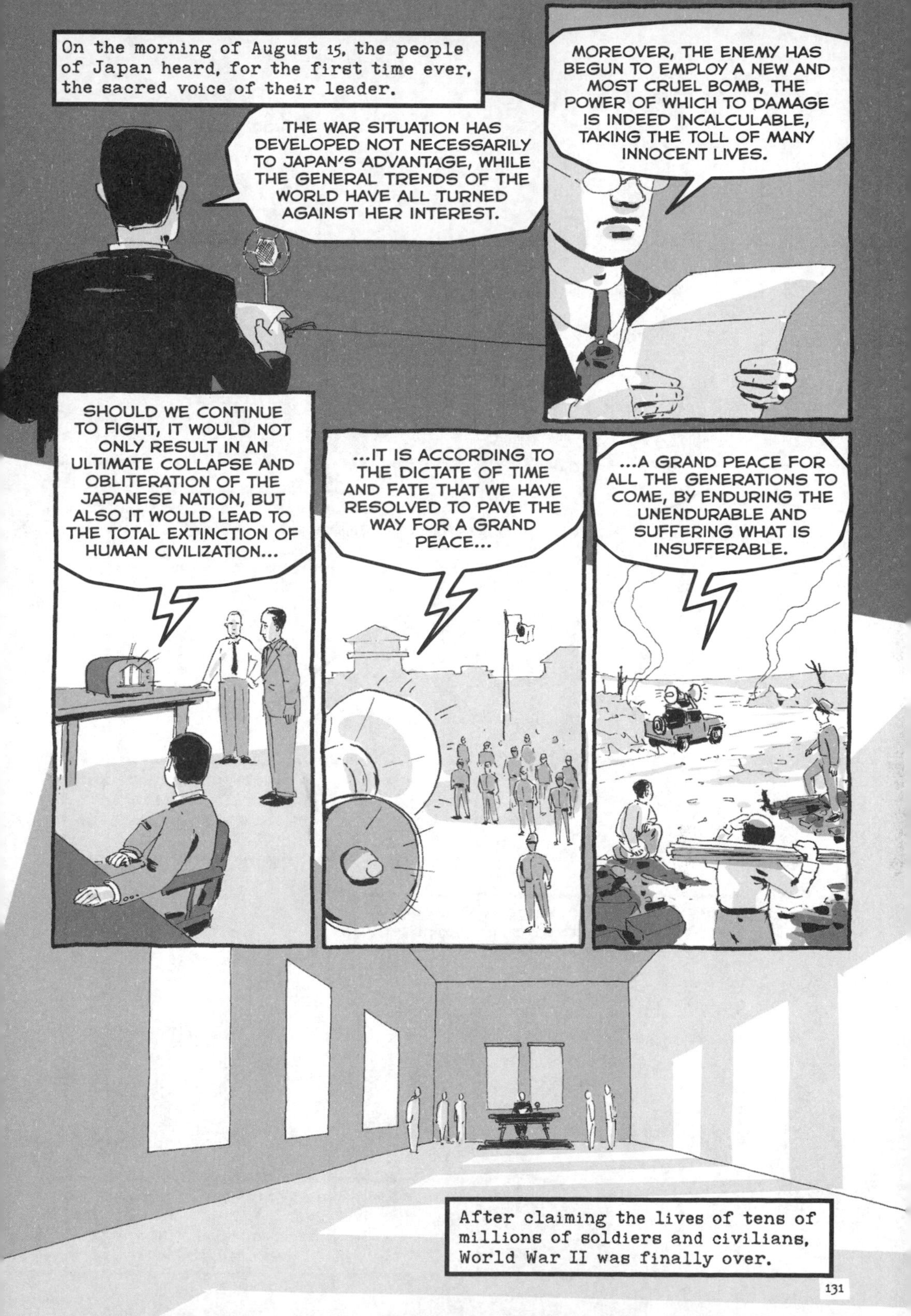
On the morning of August 15, the people of Japan heard, for the first time ever, the sacred voice of their leader.
THE WAR SITUATION HAS DEVELOPED NOT NECESSARILY TO JAPAN'S ADVANTAGE, WHILE THE GENERAL TRENDS OF THE WORLD HAVE ALL TURNED AGAINST HER INTEREST.
MOREOVER, THE ENEMY HAS BEGUN TO EMPLOY A NEW AND MOST CRUEL BOMB, THE POWER OF WHICH TO DAMAGE IS INDEED INCALCULABLE, TAKING THE TOLL OF MANY INNOCENT LIVES.
SHOULD WE CONTINUE TO FIGHT, IT WOULD NOT ONLY RESULT IN AN ULTIMATE COLLAPSE AND OBLITERATION OF THE JAPANESE NATION, BUT ALSO IT WOULD LEAD TO THE TOTAL EXTINCTION OF HUMAN CIVILIZATION...
...IT IS ACCORDING TO THE DICTATE OF TIME AND FATE THAT WE HAVE RESOLVED TO PAVE THE WAY FOR A GRAND PEACE...
...A GRAND PEACE FOR ALL THE GENERATIONS TO COME, BY ENDURING THE UNENDURABLE AND SUFFERING WHAT IS INSUFFERABLE.
After claiming the lives of tens of millions of soldiers and civilians, World War II was finally over.

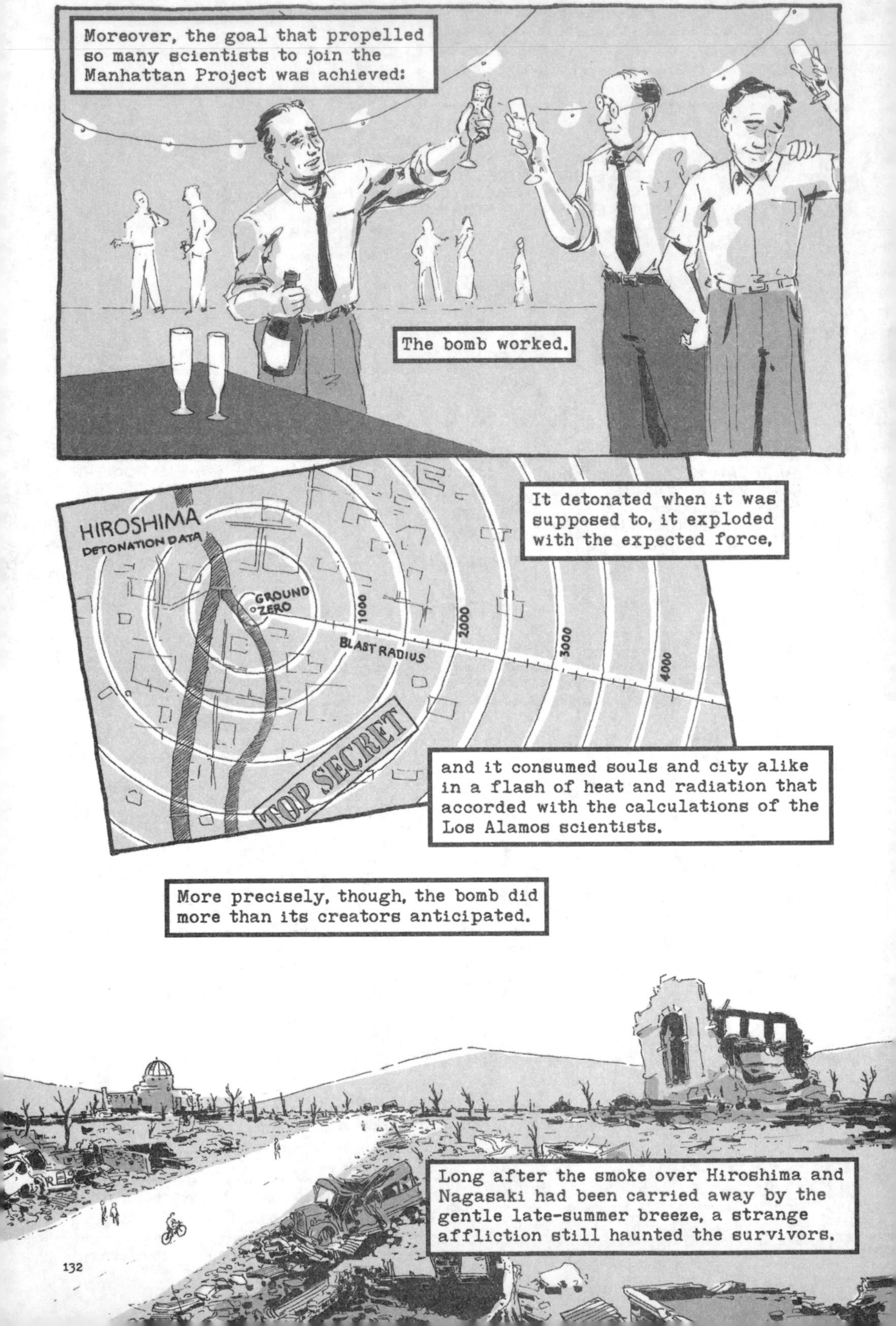
Moreover, the goal that propelled so many scientists to join the Manhattan Project was achieved:
The bomb worked.
HIROSHIMA
DETONATION DATA
GROUND ZERO
BLAST RADIUS
1000
2000
3000
4000
TOP SECRET
It detonated when it was supposed to, it exploded with the expected force,
and it consumed souls and city alike in a flash of heat and radiation that accorded with the calculations of the Los Alamos scientists.
More precisely, though, the bomb did more than its creators anticipated.
Long after the smoke over Hiroshima and Nagasaki had been carried away by the gentle late-summer breeze, a strange affliction still haunted the survivors.

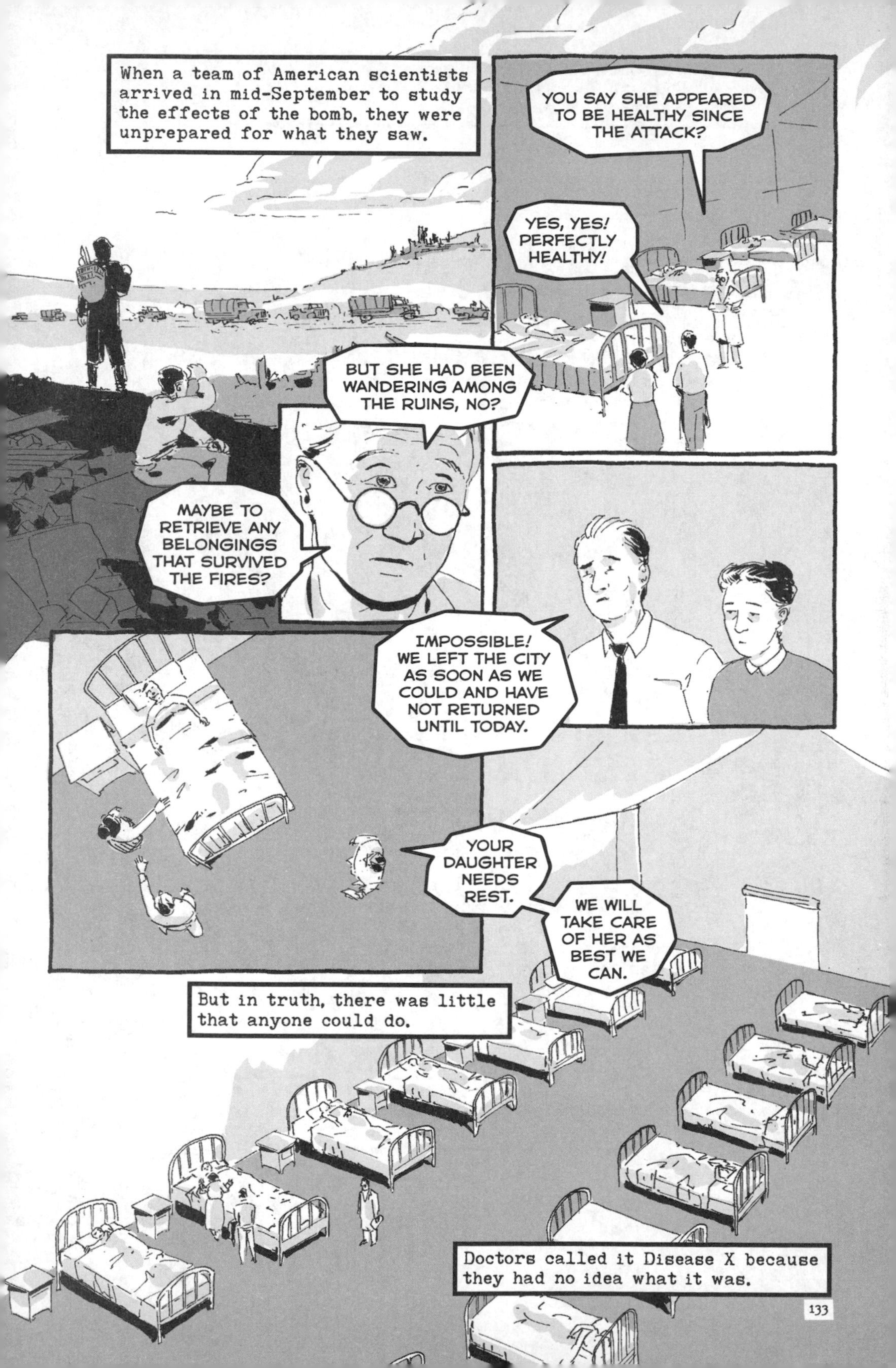
When a team of American scientists arrived in mid-September to study the effects of the bomb, they were unprepared for what they saw.
YOU SAY SHE APPEARED TO BE HEALTHY SINCE THE ATTACK?
YES, YES! PERFECTLY HEALTHY!
BUT SHE HAD BEEN WANDERING AMONG THE RUINS, NO?
MAYBE TO RETRIEVE ANY BELONGINGS THAT SURVIVED THE FIRES?
IMPOSSIBLE! WE LEFT THE CITY AS SOON AS WE COULD AND HAVE NOT RETURNED UNTIL TODAY.
YOUR DAUGHTER NEEDS REST.
WE WILL TAKE CARE OF HER AS BEST WE CAN.
But in truth, there was little that anyone could do.
Doctors called it Disease X because they had no idea what it was.

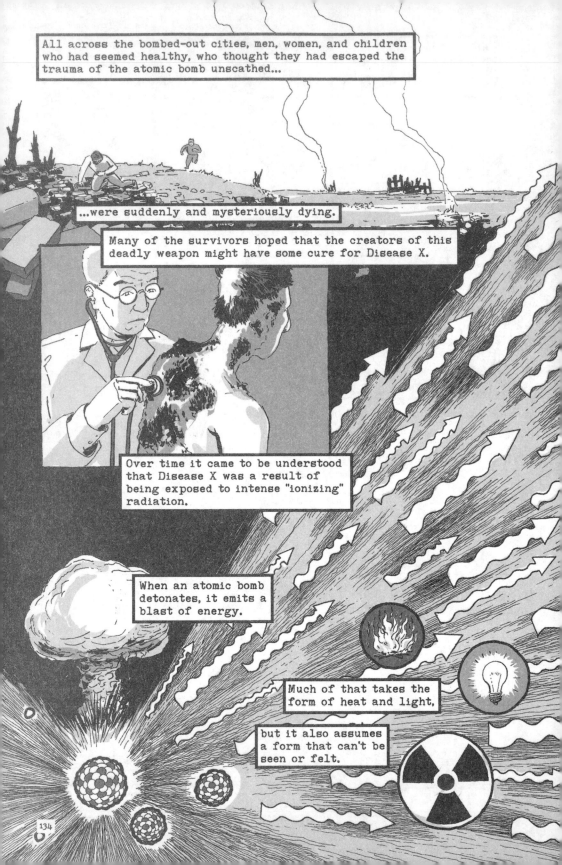

All across the bombed-out cities, men, women, and children who had seemed healthy, who thought they had escaped the trauma of the atomic bomb unscathed...

...were suddenly and mysteriously dying.

Many of the survivors hoped that the creators of this deadly weapon might have some cure for Disease X.

Over time it came to be understood that Disease X was a result of being exposed to intense "ionizing" radiation.

When an atomic bomb detonates, it emits a blast of energy.

Much of that takes the form of heat and light,

but it also assumes a form that can't be seen or felt.

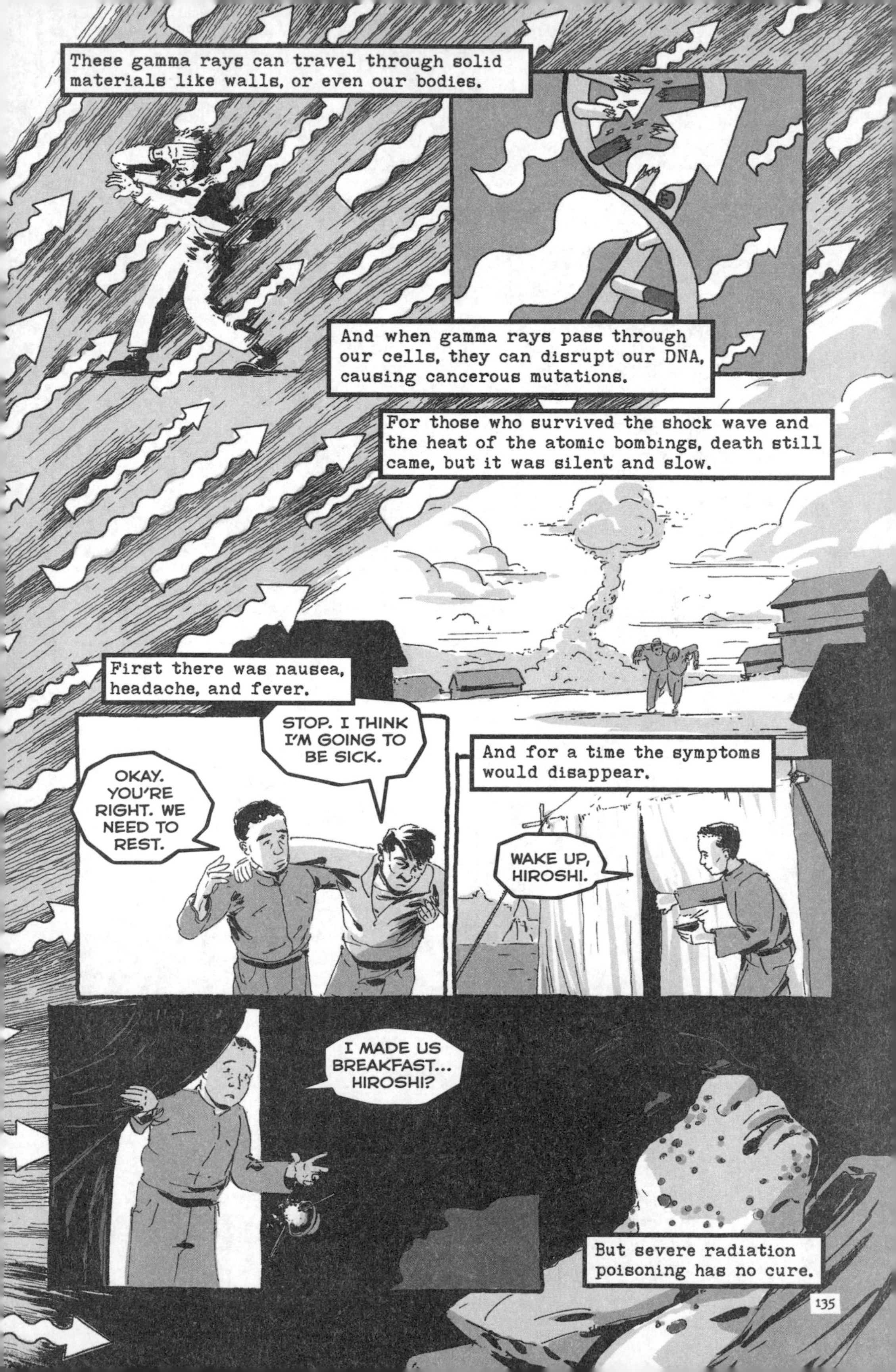
These gamma rays can travel through solid materials like walls, or even our bodies.
And when gamma rays pass through our cells, they can disrupt our DNA, causing cancerous mutations.
For those who survived the shock wave and the heat of the atomic bombings, death still came, but it was silent and slow.
First there was nausea, headache, and fever.
STOP. I THINK I'M GOING TO BE SICK.
OKAY. YOU'RE RIGHT. WE NEED TO REST.
And for a time the symptoms would disappear.
WAKE UP, HIROSHI.
I MADE US BREAKFAST... HIROSHI?
But severe radiation poisoning has no cure.

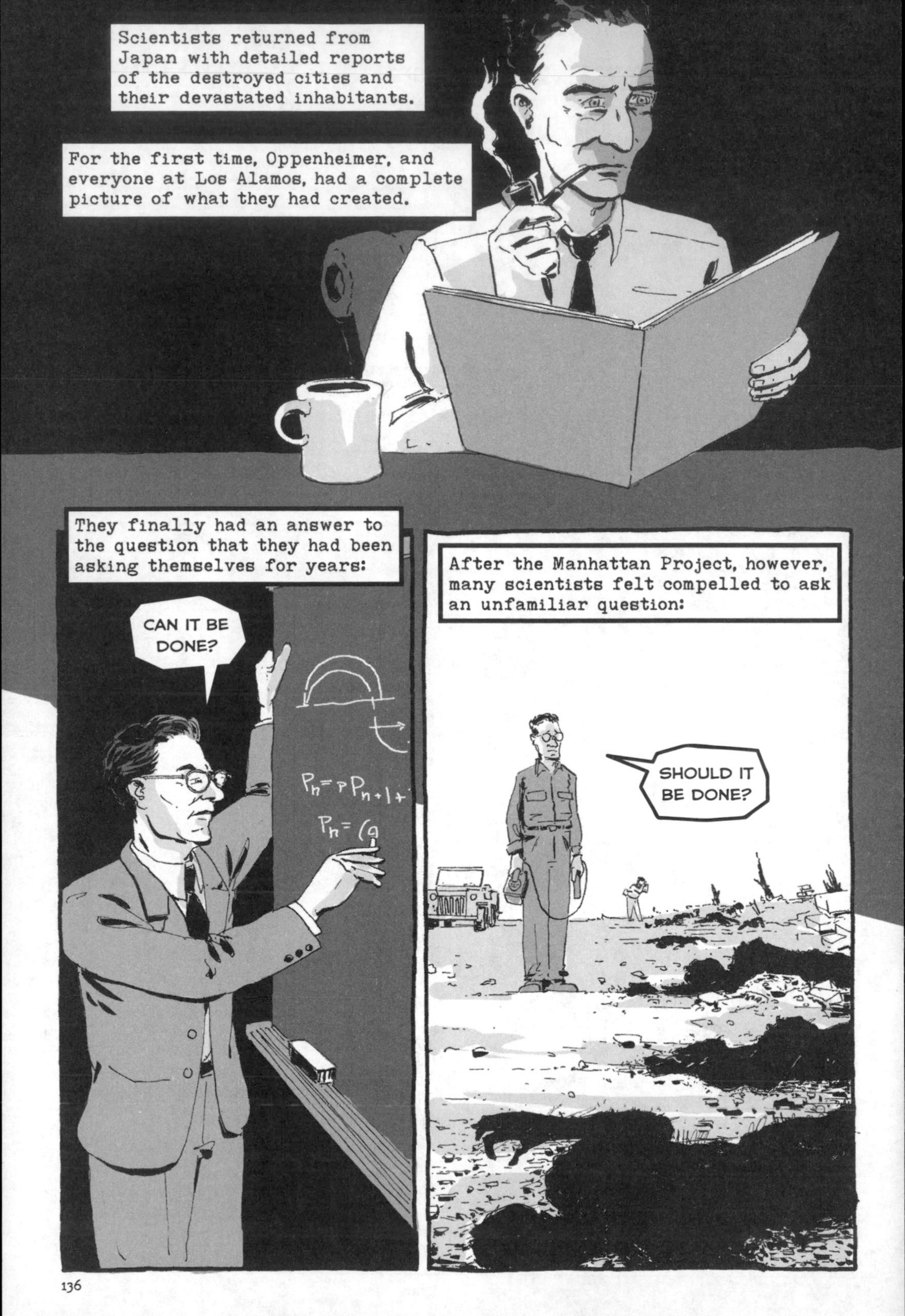
Scientists returned from Japan with detailed reports of the destroyed cities and their devastated inhabitants.
For the first time, Oppenheimer, and everyone at Los Alamos, had a complete picture of what they had created.
They finally had an answer to the question that they had been asking themselves for years:
CAN IT BE DONE?
After the Manhattan Project, however, many scientists felt compelled to ask an unfamiliar question:
SHOULD IT BE DONE?

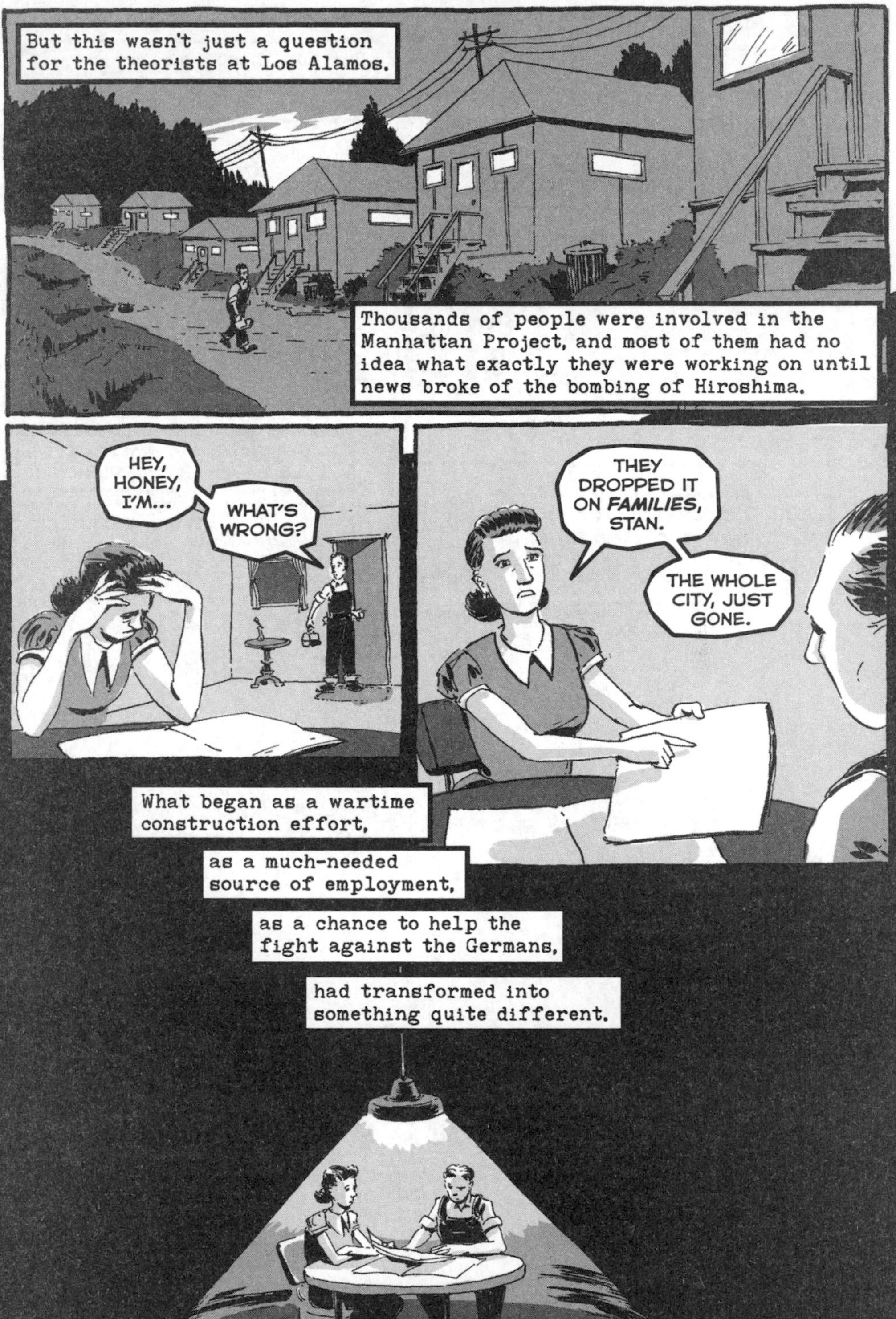
But this wasn't just a question for the theorists at Los Alamos.
Thousands of people were involved in the Manhattan Project, and most of them had no idea what exactly they were working on until news broke of the bombing of Hiroshima.
HEY, HONEY, I'M...
WHAT'S WRONG?
THEY DROPPED IT ON *FAMILIES*, STAN.
THE WHOLE CITY, JUST GONE.
What began as a wartime construction effort,
as a much-needed source of employment,
as a chance to help the fight against the Germans,
had transformed into something quite different.
Workers who were now fully informed faced a new choice.

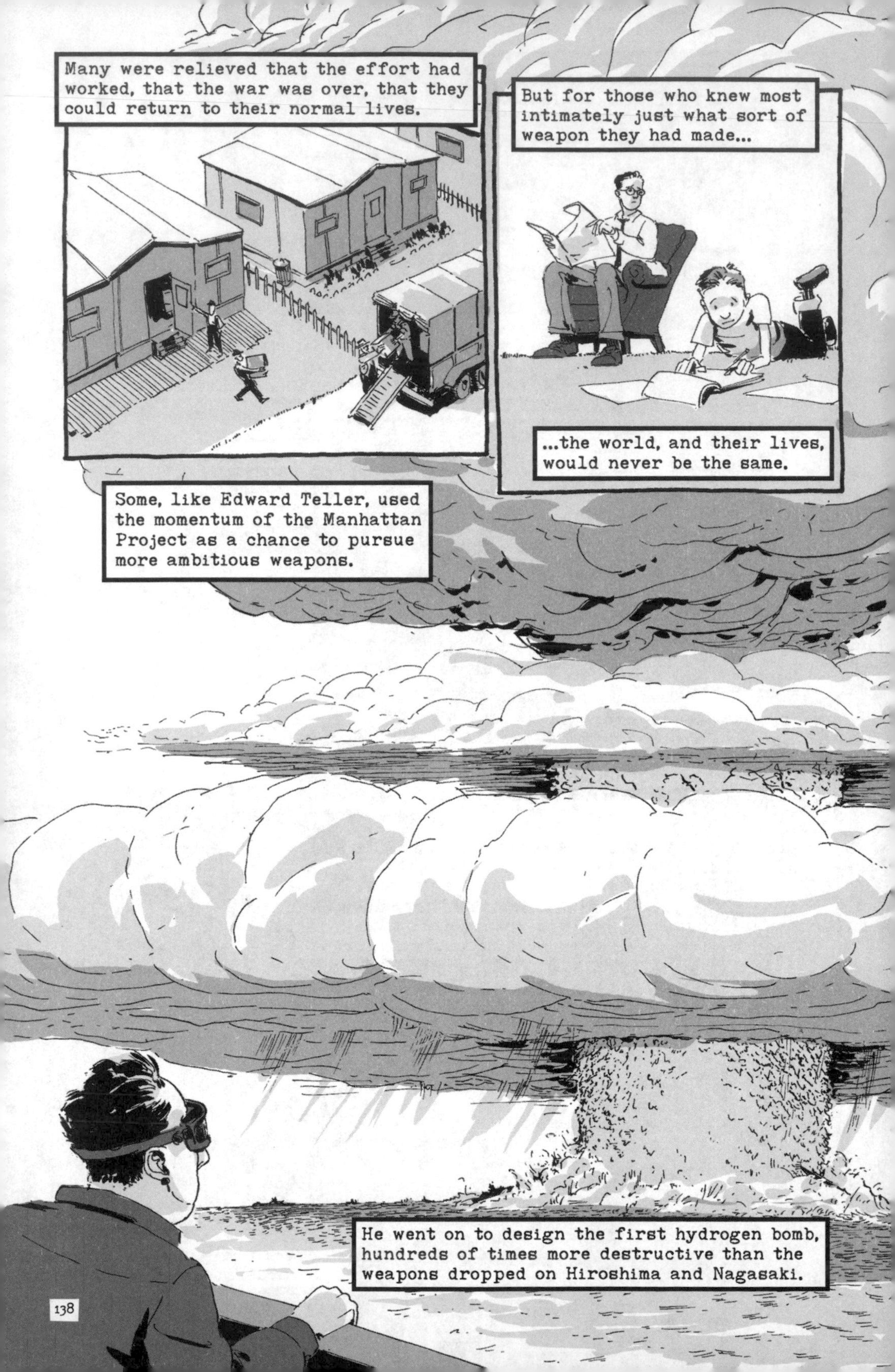
Many were relieved that the effort had worked, that the war was over, that they could return to their normal lives.
But for those who knew most intimately just what sort of weapon they had made...
...the world, and their lives, would never be the same.
Some, like Edward Teller, used the momentum of the Manhattan Project as a chance to pursue more ambitious weapons.
He went on to design the first hydrogen bomb, hundreds of times more destructive than the weapons dropped on Hiroshima and Nagasaki.

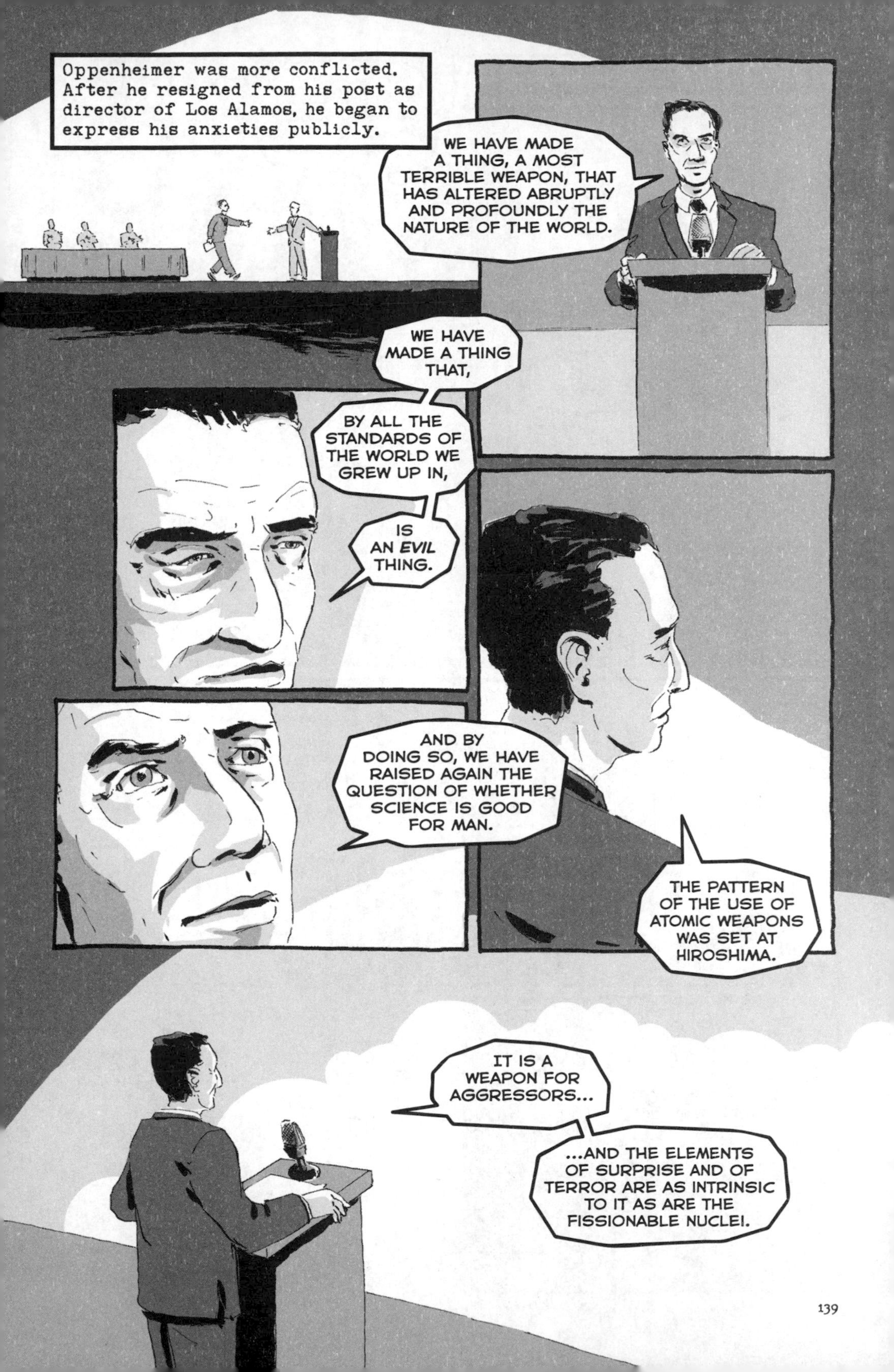
Oppenheimer was more conflicted. After he resigned from his post as director of Los Alamos, he began to express his anxieties publicly.
WE HAVE MADE A THING, A MOST TERRIBLE WEAPON, THAT HAS ALTERED ABRUPTLY AND PROFOUNDLY THE NATURE OF THE WORLD.
WE HAVE MADE A THING THAT,
BY ALL THE STANDARDS OF THE WORLD WE GREW UP IN,
IS AN *EVIL* THING.
AND BY DOING SO, WE HAVE RAISED AGAIN THE QUESTION OF WHETHER SCIENCE IS GOOD FOR MAN.
THE PATTERN OF THE USE OF ATOMIC WEAPONS WAS SET AT HIROSHIMA.
IT IS A WEAPON FOR AGGRESSORS...
...AND THE ELEMENTS OF SURPRISE AND OF TERROR ARE AS INTRINSIC TO IT AS ARE THE FISSIONABLE NUCLEI.

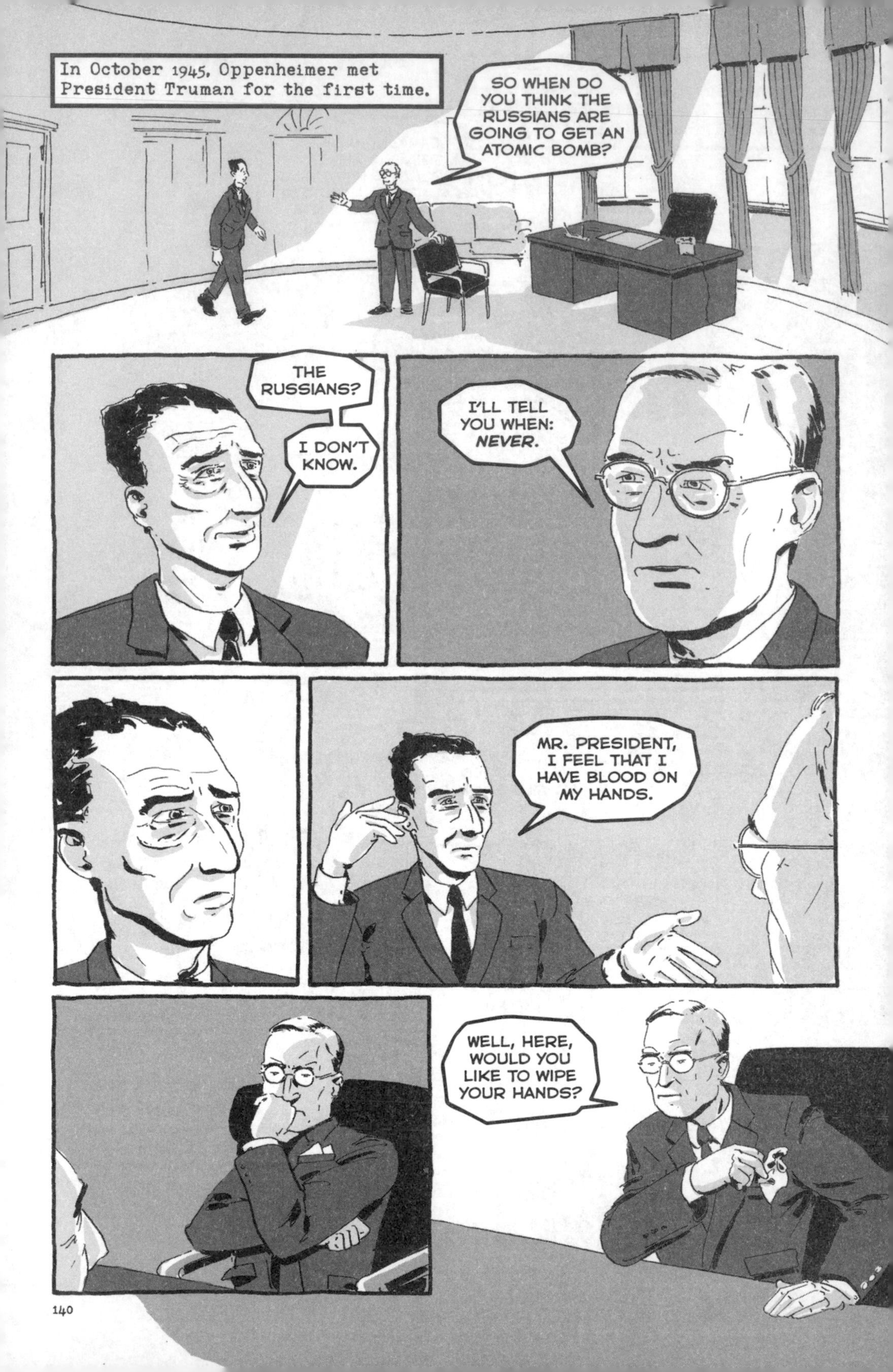
In October 1945, Oppenheimer met President Truman for the first time.
SO WHEN DO YOU THINK THE RUSSIANS ARE GOING TO GET AN ATOMIC BOMB?
THE RUSSIANS?
I DON'T KNOW.
I'LL TELL YOU WHEN: *NEVER*.
MR. PRESIDENT, I FEEL THAT I HAVE BLOOD ON MY HANDS.
WELL, HERE, WOULD YOU LIKE TO WIPE YOUR HANDS?

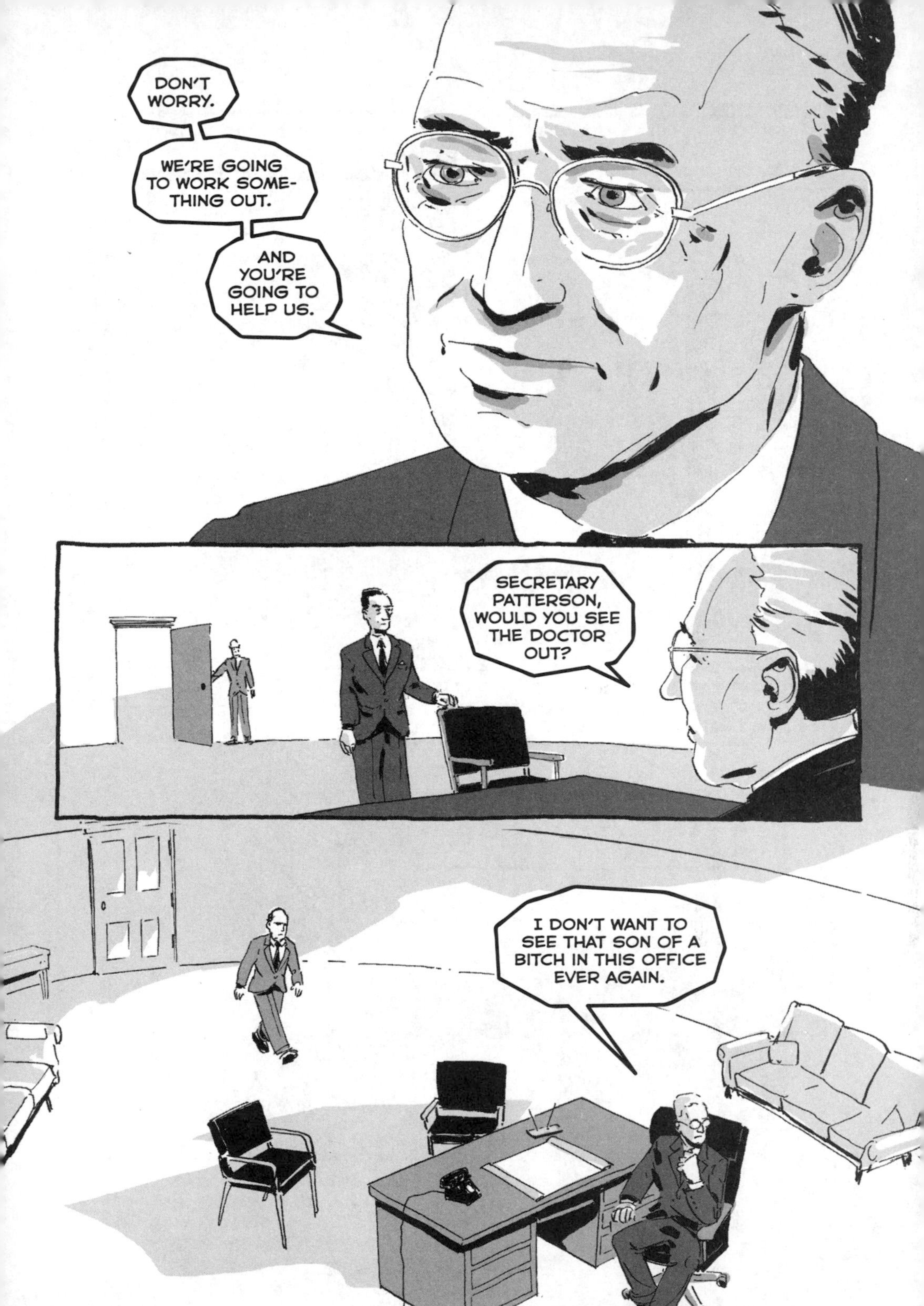
DON'T WORRY.
WE'RE GOING TO WORK SOME-THING OUT.
AND YOU'RE GOING TO HELP US.
SECRETARY PATTERSON, WOULD YOU SEE THE DOCTOR OUT?
I DON'T WANT TO SEE THAT SON OF A BITCH IN THIS OFFICE EVER AGAIN.

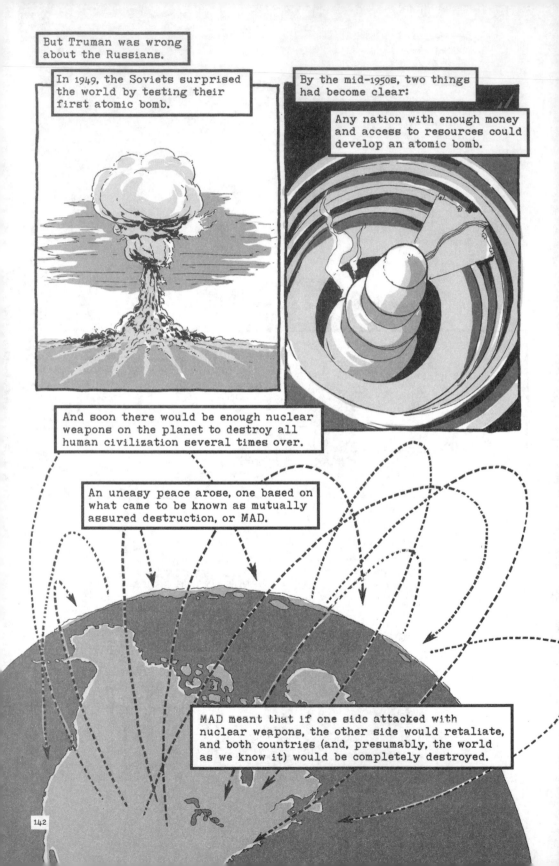

But Truman was wrong about the Russians.

In 1949, the Soviets surprised the world by testing their first atomic bomb.

By the mid-1950s, two things had become clear:

Any nation with enough money and access to resources could develop an atomic bomb.

And soon there would be enough nuclear weapons on the planet to destroy all human civilization several times over.

An uneasy peace arose, one based on what came to be known as mutually assured destruction, or MAD.

MAD meant that if one side attacked with nuclear weapons, the other side would retaliate, and both countries (and, presumably, the world as we know it) would be completely destroyed.

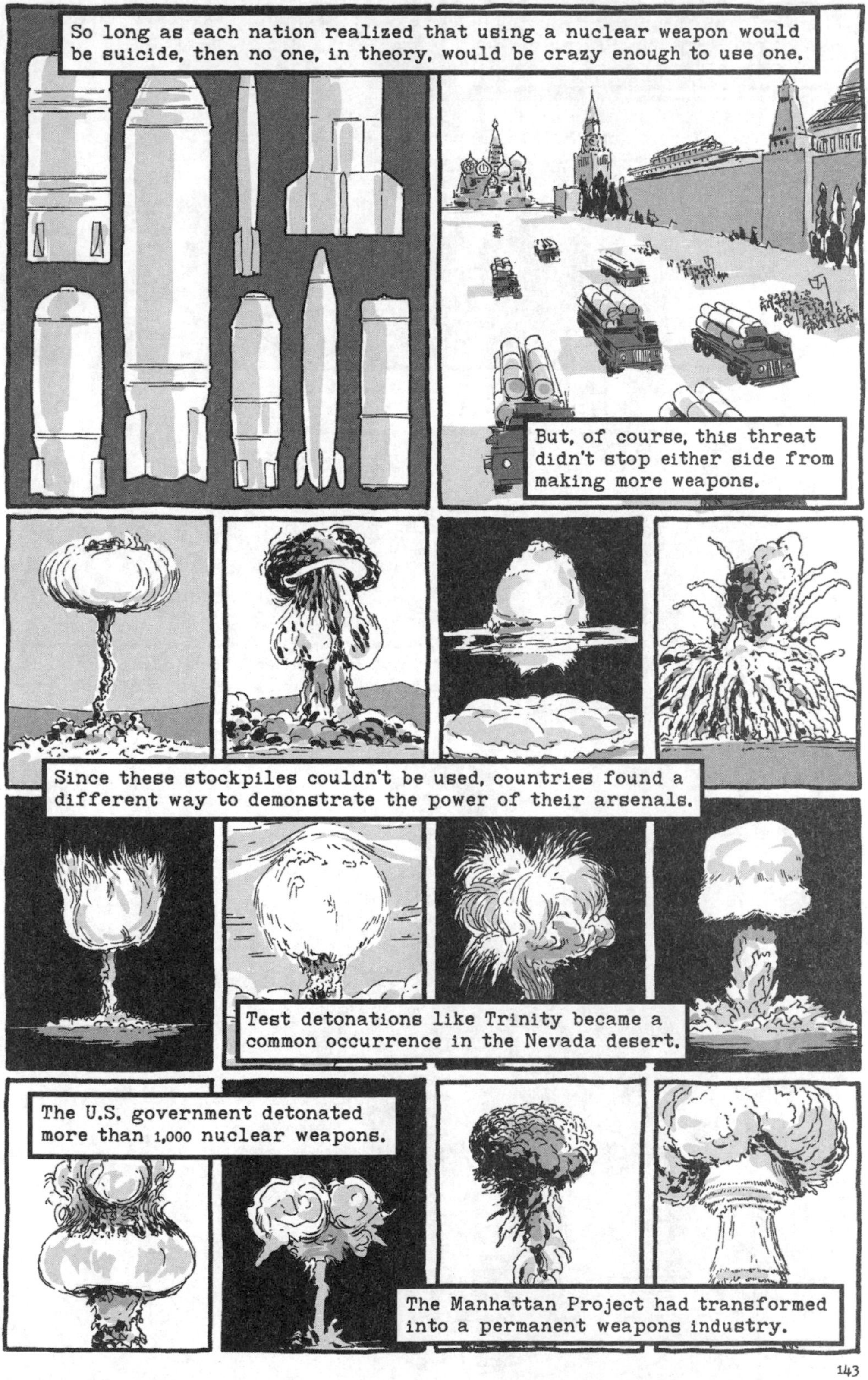
So long as each nation realized that using a nuclear weapon would be suicide, then no one, in theory, would be crazy enough to use one.
But, of course, this threat didn't stop either side from making more weapons.
Since these stockpiles couldn't be used, countries found a different way to demonstrate the power of their arsenals.
Test detonations like Trinity became a common occurrence in the Nevada desert.
The U.S. government detonated more than 1,000 nuclear weapons.
The Manhattan Project had transformed into a permanent weapons industry.

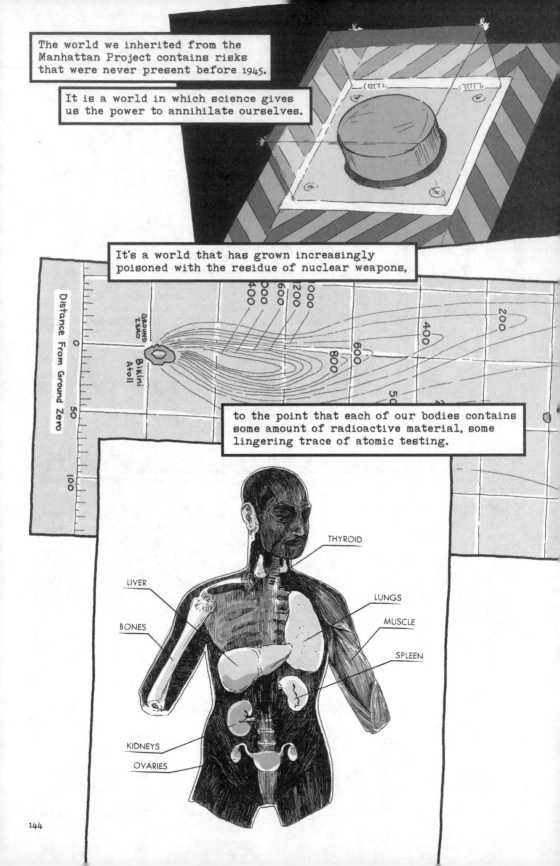

The world we inherited from the Manhattan Project contains risks that were never present before 1945.

It is a world in which science gives us the power to annihilate ourselves.

It's a world that has grown increasingly poisoned with the residue of nuclear weapons,

to the point that each of our bodies contains some amount of radioactive material, some lingering trace of atomic testing.

Distance From Ground Zero

0

50

100

GROUND ZERO

Bikini Atoll

400
1000
600
1200
1000

200

400

600

800

50

THYROID

LIVER

LUNGS

BONES

MUSCLE

SPLEEN

KIDNEYS

OVARIES

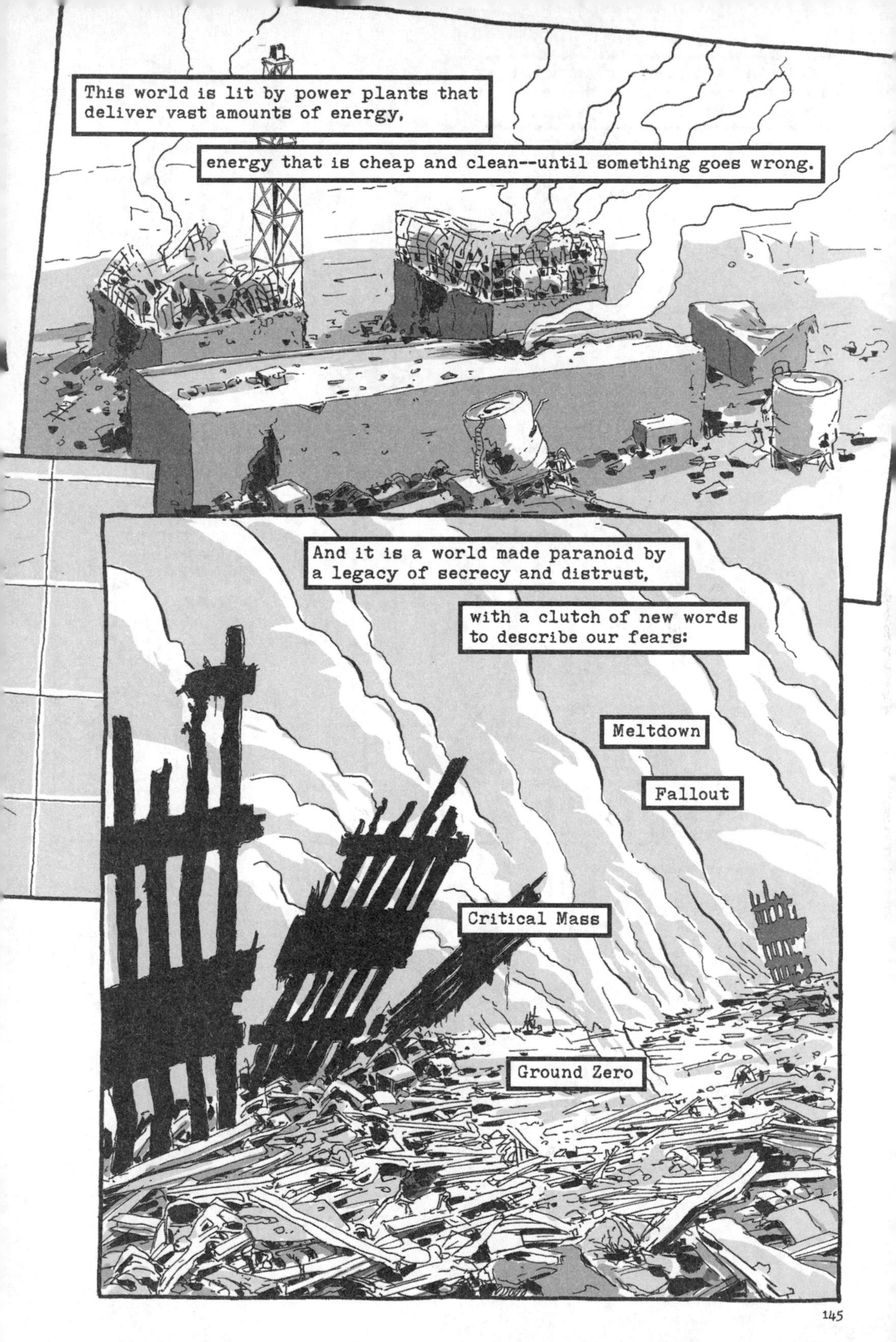
This world is lit by power plants that deliver vast amounts of energy,
energy that is cheap and clean--until something goes wrong.
And it is a world made paranoid by a legacy of secrecy and distrust,
with a clutch of new words to describe our fears:
Meltdown
Fallout
Critical Mass
Ground Zero

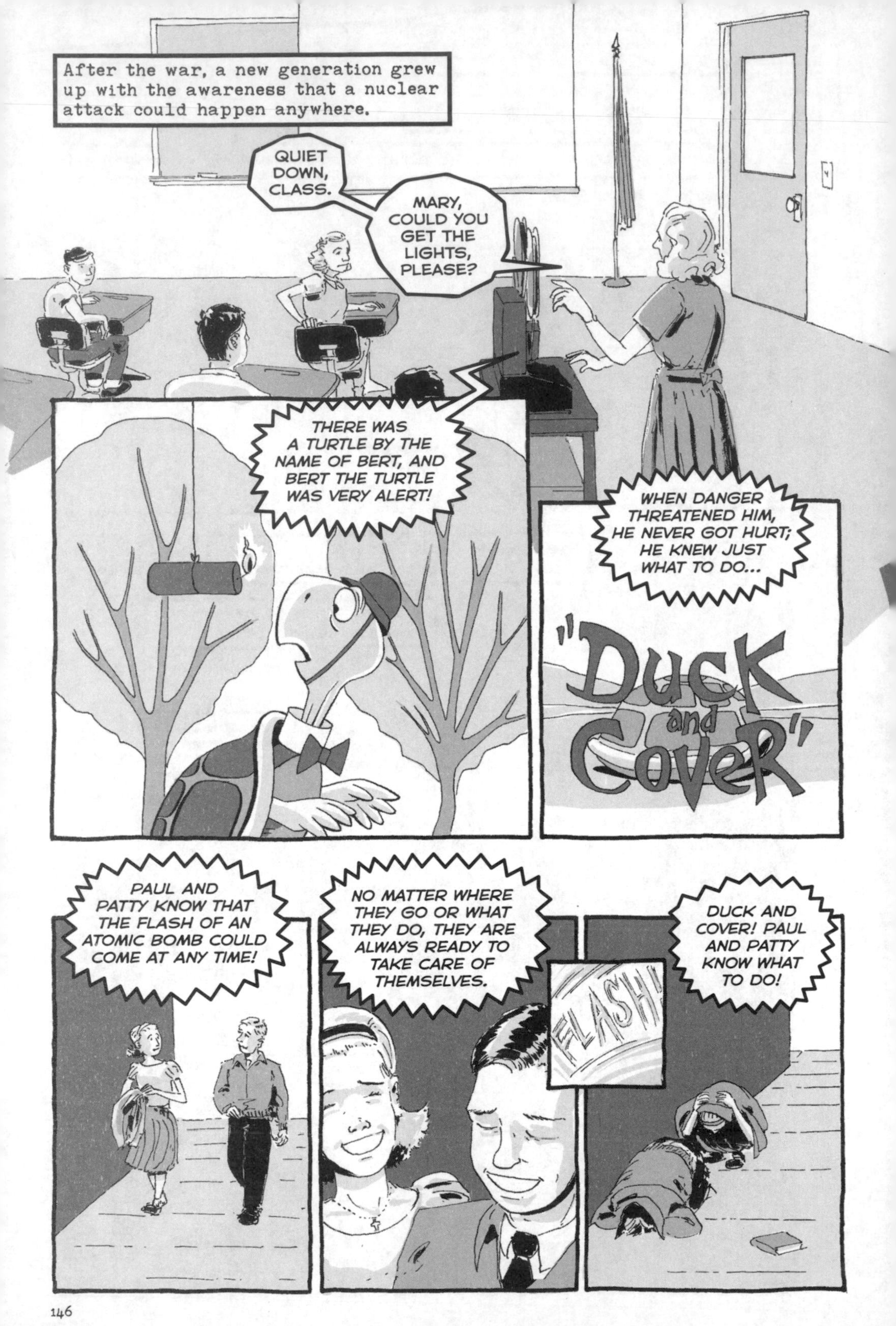
After the war, a new generation grew up with the awareness that a nuclear attack could happen anywhere.
QUIET DOWN, CLASS.
MARY, COULD YOU GET THE LIGHTS, PLEASE?
THERE WAS A TURTLE BY THE NAME OF BERT, AND BERT THE TURTLE WAS VERY ALERT!
WHEN DANGER THREATENED HIM, HE NEVER GOT HURT; HE KNEW JUST WHAT TO DO...
"DUCK and COVER"
PAUL AND PATTY KNOW THAT THE FLASH OF AN ATOMIC BOMB COULD COME AT ANY TIME!
NO MATTER WHERE THEY GO OR WHAT THEY DO, THEY ARE ALWAYS READY TO TAKE CARE OF THEMSELVES.
FLASH
DUCK AND COVER! PAUL AND PATTY KNOW WHAT TO DO!

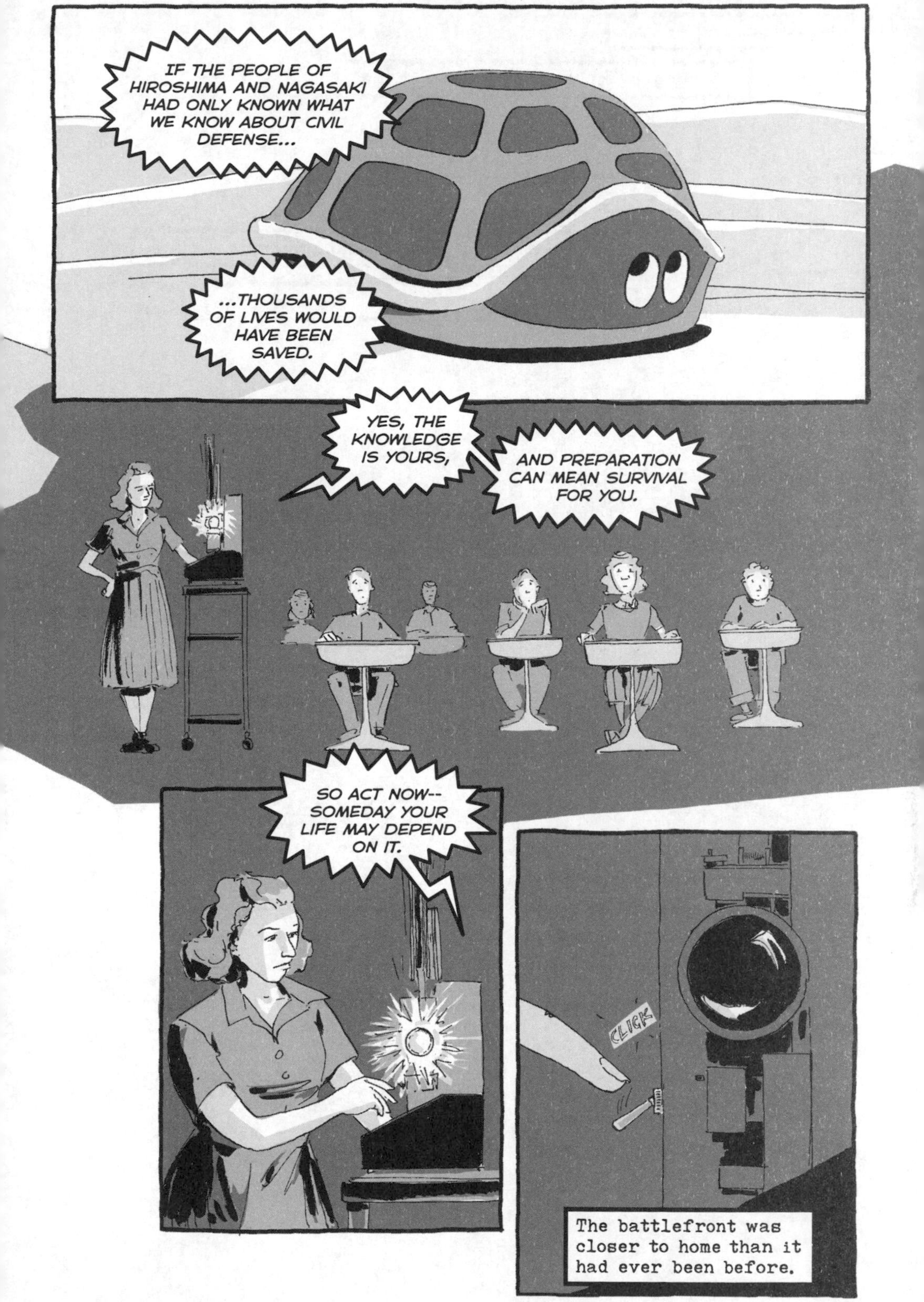
IF THE PEOPLE OF HIROSHIMA AND NAGASAKI HAD ONLY KNOWN WHAT WE KNOW ABOUT CIVIL DEFENSE...
...THOUSANDS OF LIVES WOULD HAVE BEEN SAVED.
YES, THE KNOWLEDGE IS YOURS,
AND PREPARATION CAN MEAN SURVIVAL FOR YOU.
SO ACT NOW-- SOMEDAY YOUR LIFE MAY DEPEND ON IT.
CLICK
The battlefront was closer to home than it had ever been before.

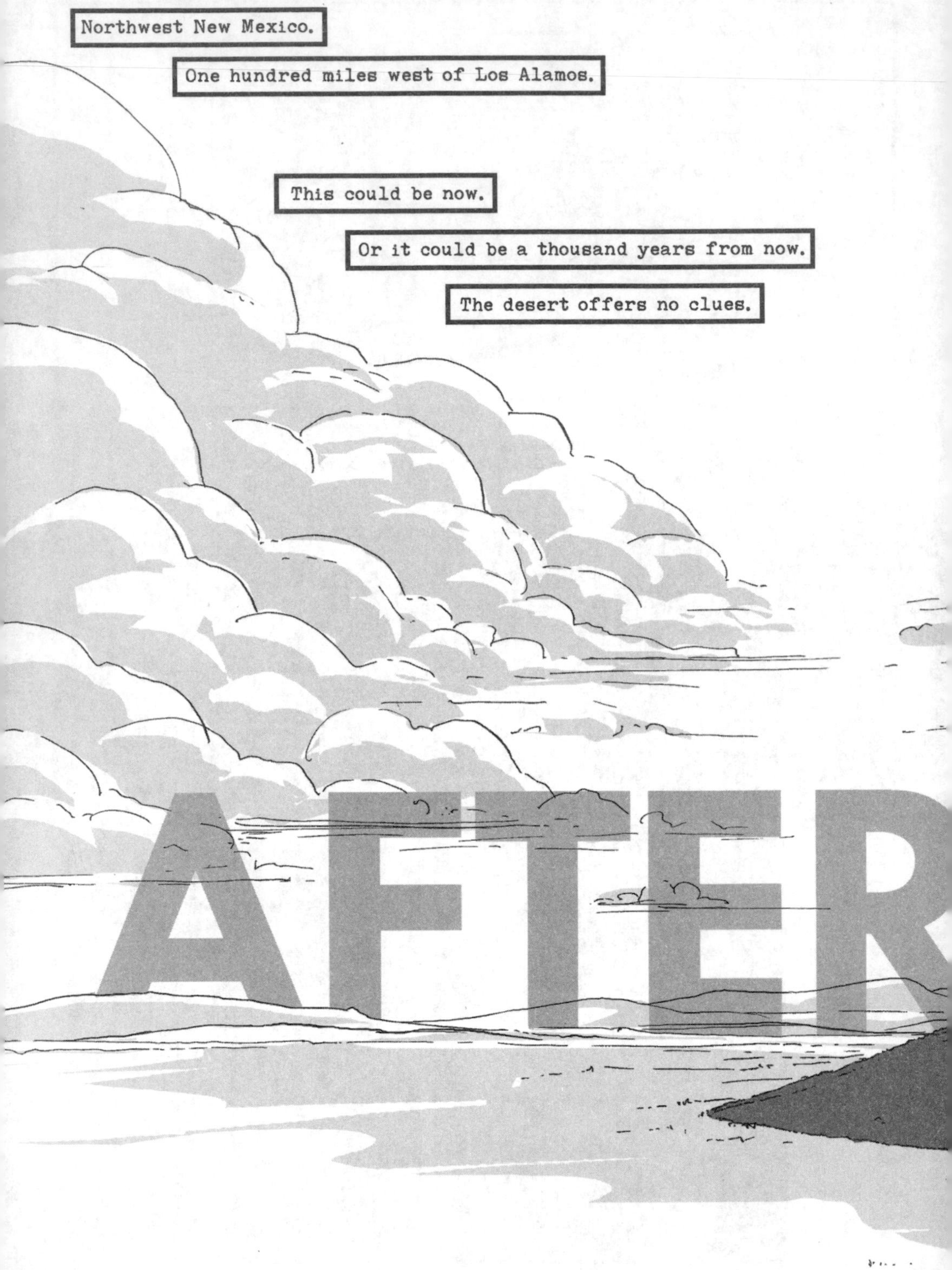
Northwest New Mexico.
One hundred miles west of Los Alamos.
This could be now.
Or it could be a thousand years from now.
The desert offers no clues.
AFTER

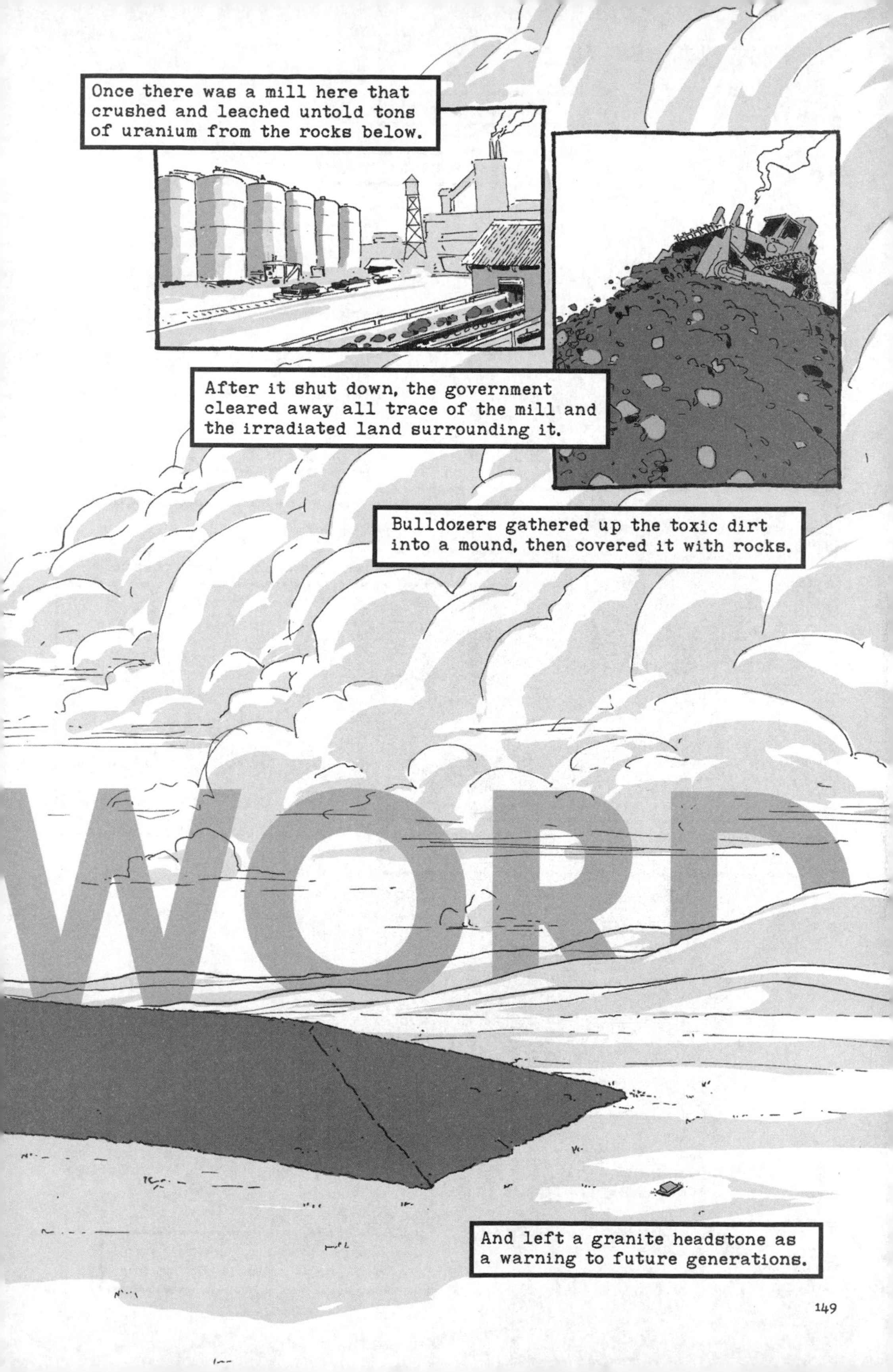
Once there was a mill here that crushed and leached untold tons of uranium from the rocks below.
After it shut down, the government cleared away all trace of the mill and the irradiated land surrounding it.
Bulldozers gathered up the toxic dirt into a mound, then covered it with rocks.
WORD
And left a granite headstone as a warning to future generations.

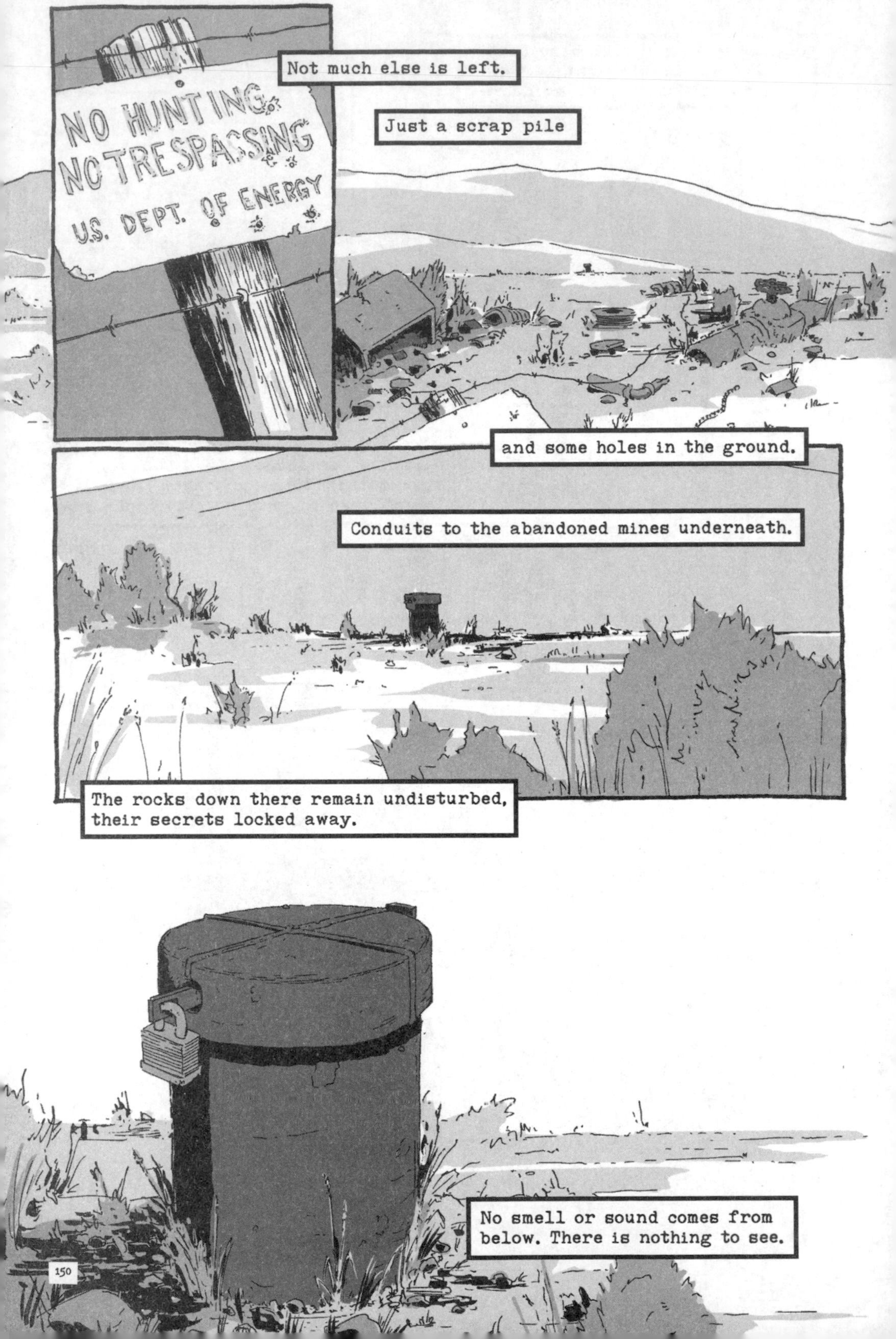
NO HUNTING
NO TRESPASSING
U.S. DEPT. OF ENERGY
Not much else is left.
Just a scrap pile
and some holes in the ground.
Conduits to the abandoned mines underneath.
The rocks down there remain undisturbed, their secrets locked away.
No smell or sound comes from below. There is nothing to see.

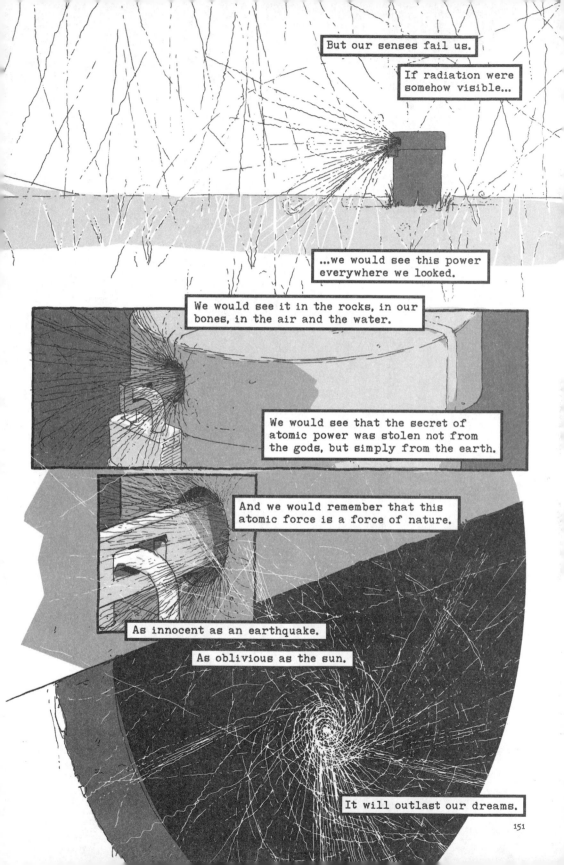

But our senses fail us.

If radiation were somehow visible...

...we would see this power everywhere we looked.

We would see it in the rocks, in our bones, in the air and the water.

We would see that the secret of atomic power was stolen not from the gods, but simply from the earth.

And we would remember that this atomic force is a force of nature.

As innocent as an earthquake.

As oblivious as the sun.

It will outlast our dreams.

Author's Note

For the most part, this is a work of history. Which means that, for the most part, the dialog from the principle characters in this book is taken from written records. When that was impossible, I introduced language that hews closely to what I have learned of these characters over the course of my research. To make this book as accurate as possible, I have relied heavily on the following works:

Bird, Kai, and Martin J. Sherwin. *American Prometheus: The Triumph and Tragedy of J. Robert Oppenheimer.* New York: Vintage, 2006.

Coster-Mullen, John. *Atom Bombs: The Top Secret Inside Story of Little Boy and Fat Man.* Self-published, 2002.

Fermi, Rachel, and Esther Samra. *Picturing the Bomb: Photographs from the Secret World of the Manhattan Project.* New York: Harry N. Abrams, 1995.

Gibson, Toni Michnovicz, and Jon Michnovicz. *Los Alamos: 1944–1947.* Charleston, South Carolina: Arcadia Publishing, 2005.

Goldstein, Donald M., Katherine V. Dillon, and J. Michael Wenger. *Rain of Ruin: A Photographic History of Hiroshima and Nagasaki.* Washington: Potomac Books, Inc., 1999.

Gordin, Michael D. *Five Days in August: How World War II Became a Nuclear War.* Princeton, New Jersey: Princeton University Press, 2007.

Groves, Leslie R. *Now It Can Be Told: The Story of the Manhattan Project.* New York: Da Capo Press, 1983.

Joseph, Timothy. *Historic Photos of the Manhattan Project.* Nashville, Tennessee: Turner Publishing, 2009.

Masters, Dexter, and Katharine Way, eds. *One World or None.* New York: McGraw-Hill Book Co., 1946.

O'Donnell, Joe. *Japan 1945: A U.S. Marine's Photographs from Ground Zero.* Nashville, Tennessee: Vanderbilt University Press, 2005.

Oppenheimer, J. Robert. "Atomic Weapons," *Proceedings of the American Philosophical Society,* Vol. 90, No. 1 (Jan. 1946), pp. 7–10.

Rhodes, Richard. *The Making of the Atomic Bomb.* New York: Simon & Schuster, 1986.

Rotter, Andrew J. *Hiroshima: The World's Bomb.* New York: Oxford University Press, 2008.

Serber, Robert. *The Los Alamos Primer.* Berkeley: University of California Press, 1992.

Spitzer, Abe. Personal Diary of Abe Spitzer. Sgt. Abe Spitzer Collection. Manhattan Project Heritage Preservation Association, Inc. Accessed June 9, 2011. http://www.mphpa.org/classic/COLLECTIONS/CG-ASPI/01/Pages/ASPI_Gallery_01.htm.

For a Closer Look...

One hundred and fifty pages are far too few to encompass the history of the atomic bomb. There are countless stories to tell, an infinity of details that should be remembered. If you're curious to learn more, here are some books, films, and websites to consider:

Hersey, John. *Hiroshima*. New York: Vintage, 1989.

Kelly, Cynthia C., ed. *The Manhattan Project*. New York: Black Dog & Leventhal Publishers, 2007.

Barefoot Gen. Directed by Mamoru Shinzaki and Toshio Hirata. Long Beach, California: Geneon USA, 2006.

The Day After Trinity. Directed by Jon Else. Chatsworth, California: Image Entertainment, 1981.

The Atomic Archive: http://www.atomicarchive.com.

The Atomic Heritage Foundation: http://www.atomicheritage.org/mediawiki.

Acknowledgments

This book was conceived in an office and completed in a studio, but there were many detours along the way, from the belly of a nuclear reactor to the dunes of a radioactive desert. I must thank Thomas LeBien for making sure I never wandered too far off course and Antonio Iannarone for helping me lose myself whenever possible. This book would have been far less than what it is if it weren't for the careful eye of Kellan Cummings and the keen mind of Katie Ryder. For their help in improving the scientific and historical accuracy of the text, I must acknowledge Cynthia Kelly, Harold Agnew, George Cowan, and Michael Lucibella. Thanks as well to Joanna Fiduccia, Philip Marino, Soma Wingelaar, and Matthew Phelan. I am ever grateful to Peter Stansky and Paula Findlen for encouraging me down this path. Finally, I owe an unquantifiable debt of gratitude to Mary Fetter, Karen Leigh, and Darrell Worm. I dedicate this book to Jan and Stan Worm.